W0227504

Lecture Notes of the Institute for Computer Sciences, Social Informatics and Telecommunications Engineering   103

Emma Hart   Jon Timmis   Paul Mitchell
Takadash Nakano   Foad Dabiri (Eds.)

Bio-Inspired Model of Network, Information, and Computing Systems

6th International ICST Conference, BIONETICS 2011
York, UK, December 5-6, 2011
Revised Selected Papers

 Springer

Volume Editors

Emma Hart
Napier University, School of Computing
Edinburgh, EH10 5DT, UK
E-mail: e.hart@napier.ac.uk

Jon Timmis
University of York, Department of Computer Science
and Department of Electronics, Heslington, York YO10 5DD, UK
E-mail: jtimmis@cs.york.ac.uk

Paul Mitchell
University of York, Department of Computer Science
and Department of Electronics, Heslington, York. YO10 5DD, UK
E-mail: pdm106@ohm.york.ac.uk

Takadash Nakano
Frontier Research Base for Global Young Researchers
Graduate School of Engineering, Osaka University
Osaka 565-0871, Japan
E-mail: tnakano@wakate.frc.eng.osaka-u.ac.jp

Foad Dabiri
University of California, Los Angeles, CA 90095, USA
E-mail: dabri@cs.ucla.edu

ISSN 1867-8211   e-ISSN 1867-822X
ISBN 978-3-642-32710-0   e-ISBN 978-3-642-32711-7
DOI 10.1007/978-3-642-32711-7

Springer Heidelberg Dordrecht London New York

Library of Congress Control Number: 2012944179

CR Subject Classification (1998): C.2, H.4, I.2.9, F.2.2, I.2.11

Typesetting: Camera-ready by author, data conversion by Scientific Publishing Services, Chennai, India

Printed on acid-free paper

Springer is part of Springer Science+Business Media (www.springer.com)

Preface

Bionetics 2011 was the 6th International ICST Conference on Bio-Inspired Models of Network, Information and Computing Systems, which we were proud to host in the beautiful city of York, UK, which combines a colorful history of Roman and Viking rule with modern science and research developments. This year's conference aimed to continue the tradition of providing a world-leading and unique opportunity for bringing together researchers and practitioners from diverse disciplines who seek the understanding of the fundamental principles and design strategies in biological systems and leverage those understandings to build bio-inspired systems.

Bionetics 2011 opened with a keynote talk by Jeremy Pitt, Reader in Intelligent Systems from Imperial College London. Jeremy's primary research interest is in the science, technology and application of multi-agent systems, especially in communications. Recently, he has been applying socio-economic principles to examine self-organizing institutions. A second keynote was given by Dario Floreano. In 2000 Prof. Floreano was awarded the first Swiss National Science Foundation professorship in bio-inspired robotics at EPFL. In 2005 he was appointed associate EPFL professor and established the Laboratory of Intelligent Systems. In 2011 he was appointed full EPFL professor and became director of the newly established Swiss National Center of Competence in Robotics.

Papers were accepted into two categories; full paper and work-in-progress. Full papers describe significant advances in the bionetics field, while work-in-progress papers present an opportunity for authors to discuss breaking research that is currently being evaluated. Seven full papers were accepted, covering topics ranging from robotic coordination to attack detection in peer-to-peer networks. A further seven work-in-progress papers were inspired by biological mechanisms including evolution, flocking and artificial immune systems. In addition to the main conference track, the conference hosted a parallel stream in the field of "Nano-scale Communication and Networking", chaired by Tadashi Nakano and Michael Moore.

We were delighted to host a PhD forum at Bionetics 2011 to give an opportunity to young researchers to present their work, whatever the stage of their PhD, in a friendly and supportive environment. Researchers attending the forum benefited greatly from the advice given by Prof. A.E. Eiben from the Vrije University, Amsterdam, who has many years' experience of supervising students in the area of biologically inspired computing.

The Bionetics 2011 team would like to thank all the sponsors, including ICST, Create-Net, EAI, and the AWARENESS coordination action, the latter generously provided support for a number of student bursaries and the keynote speaker for the PhD forum. In addition, we extend our thanks to all those who helped with the conference, including the many people who reviewed papers in the Program Committees, managed publicity and the website and helped in any other way, big or small.

November 2011   Emma Hart
                                                                          Jon Timmis

Organization

General Chair

Jon Timmis
University of York, UK
jtimmis@cs.york.ac.uk

Technical PC Chair

Emma Hart
Edinburgh Napier University, UK
e.hart@napier.ac.uk

Steering Committee

Imrich Chlamtac
Create-Net and University of Trento, Italy

Tatsuya Suda
University of California, Irvine, USA

Falko Dressler
University of Erlangen, Germany

Iacopo Carreras
Create-Net, Italy

Tadashi Nakano
Osaka University, Japan

Jun Suzuki
University of Massachusetts, Boston, USA

Publication Chair

Foad Dabiri
University of California, Los Angeles, USA
dabiri@cs.ucla.edu

Special Sessions Chair

Paul Mitchell
University of York, UK
pdm106@ohm.york.ac.uk

Publicity Chair

Mark Read
University of York, UK
mnr101@ohm.york.ac.uk

Web Chair

James Hilder
University of York, UK
jah128@ohm.york.ac.uk

Table of Contents

NANO Special Session

PhD Forum

The Price of Evolution
in Incremental Network Design
(The Case of Ring Networks)*

Saeideh Bakhshi and Constantine Dovrolis

College of Computing, Georgia Tech
{saeideh,constantine}@gatech.edu

Abstract. As it also happens in nature, technological networks typically
evolve in an incremental manner, instead of being optimally designed. This
evolutionary process is driven by changes in the underlying parameters
and constraints (the "environment") and it typically aims to minimize the
modification cost after each change in the environment. In this paper, we
first formulate the incremental network design approach and compare that
with the more traditional optimized design approach in which the objective is to minimize the total network cost. We evaluate the cost overhead
and evolvability of incremental design under two network expansion models (random and gradual), focusing on the simpler case of "ring" networks.
We find that even though incremental design has some cost overhead, that
overhead does not increase as the network grows. Also, it is less costly to
evolve an existing network than to design it from scratch as long as the
network expansion factor is less than a critical value.

1 Introduction

Complex technological systems, such as transportation and communication networks, manufacturing processes, microprocessors and computer operating systems, are rarely designed "from scratch." Instead, they are often subject to an
evolutionary process in which existing designs are incrementally modified every
time a new phase or version of the system is needed. There are numerous examples and they span every engineering discipline. To mention one of them, consider
a wide-area communication network that expands over time to reach new locations, increasing its capacity depending on the offered load, and occasionally
providing new services.

It is remarkable that even though the optimized design of complex systems
and networks has been studied in depth for several decades, the literature rarely
considers that the design process is often incremental (or evolutionary). Instead,
it is typically assumed that the system is designed *tabula rasa*, in a "clean-
slate" manner. The corresponding design problems are typically formulated as
optimization problems with multiple constraints, none of which is imposed by

---

* This research was supported by the National Science Foundation under Grant No.
  0831848.

E. Hart et al. (Eds.): BIONETICS 2011, LNICST 103, pp. 1–15, 2012.

an earlier design however. In the few previous studies that considered evolving networks (see Section 4), the focus was on the design algorithms and the corresponding optimization problems, rather than to examine the pros and cons of an incremental design relative the corresponding optimized design.

In this paper, we attempt a first investigation of the following fundamental question: *how does an incremental design compare to an optimized design, when both designs provide the same function?* Even though we believe that the previous question is relevant to the study of complex systems in general, we choose to focus on a rather narrow design problem and the engineering domain of our expertise, namely *the topological design of communication "ring" networks.* The ring topology is often used in practice mostly in metropolitan or regional networks [13]. In that context, the objective of the design process is to create a ring network that interconnects a given set of locations aiming to minimize a cost-related objective function. We further limit (admittedly in a very simplified manner) how the environment changes with time: the set of interconnected network locations expands by one or more nodes at each time step.

The previous ring topology design problem allows us to examine several interesting questions about incremental versus optimized designs in a precise and quantitative framework. How can we mathematically formulate the incremental design problem, and how is that formulation different from the more traditional optimized design problem? How does an incremental design compare, in terms of cost with the corresponding clean-slate design that provides the same function? How costly is it to modify an existing design, relative to the cost of re-designing the system from scratch? When is it better to abandon incremental changes on an existing system and start from scratch? What is the *"price of evolution"*, i.e., the cost overhead of an incremental design relative to the corresponding optimized design? What is the role of the pace at which the environment changes with time? Does the incremental design process perform better when the environment varies in certain non-random ways?

In Section 2, we present a formulation for the incremental design problem in which the objective is to *minimize the modification cost relative to the existing network.* We compare that formulation with an optimized design problem in which the objective is to minimize the total network cost. In Section 3, we derive expressions for the evolvability and cost overhead of the incremental design process. We also compare *random expansion* with *gradual expansion.* We review the related work in Section 4 and conclude in Section 5. In a follow-up paper we will expand this study in the case of general mesh networks.

2   Framework and Metrics

In this section, we present mathematical formulations for the clean-slate and incremental design problems. Even though these formulations are quite general, in the rest of the paper we apply them in the context of topology design for communication networks. As in any design problem, there is a desired function (e.g., construct a communication network to connect a given set of locations), some design elements (e.g., routers, wide-area links), certain constraints (e.g., related to reliability or maximum propagation delay) as well as an objective (e.g., minimize the total cost of the required design elements). The design process aims to use the appropriate elements so that we achieve the desired function, while satisfying the constraints and meeting a given objective. It is often assumed that this process is conducted only once in an otherwise static environment. Here, we consider the case that the design takes place in a *dynamic environment*. In the context of communication networks, the network may gradually expand to new locations, the cost of design elements may fluctuate, or the constraints may become more stringent from time to time. We consider a discrete-time model and we refer to the k'th time epoch as the k'th *environment*. At a given environment k, all inputs of the design problem are known and constant.

How can we design a communication network in a dynamic environment? We identify two fundamentally different approaches. In the *clean-slate approach* we aim *to minimize in every environment the total cost of the network* subject to the given constraints. We refer to the resulting network in each environment as *optimized*. In the *incremental approach* we aim instead to *minimize the modification cost relative to the network of the previous environment*, again subject to the given constraints; we refer to the resulting network as *evolved*.

More rigorously, let $\mathcal{N}(k)$ be the set of *acceptable networks* at environment k, i.e., networks that provide the desired function and meet the given constraints at environment k. The cost of a particular network $N \in \mathcal{N}(k)$ is $C(N)$. $C(N)$ is the sum of the costs of all design elements in N. We assume that there are no other costs associated with N; for instance, there is no monetary cost to compute the design or to interconnect its elements.

In clean-slate design, the objective is to identify an acceptable network $N_{opt}(k)$ from the set $\mathcal{N}(k)$ that has the minimum cost $C_{opt}(k)$ at environment k,

$$C_{opt}(k) \equiv C(N_{opt}(k)), \quad N_{opt}(k) \equiv \arg \min_{N \in \mathcal{N}(k)} C(N). \tag{1}$$

We refer to $N_{opt}(k)$ as the *optimized* network at environment k. If the optimized network is not unique, we break ties with secondary objectives (for instance, minimize the total number of links). Most network design problems are computationally intractable (NP-hard), and so they are often solved heuristically, approximating the previous optimization objective. This is what we also do in Section 3. For this reason, we do not refer to $N_{opt}(k)$ as *optimal* but as *optimized*. The former would be the actual solution to the previous problem if we could compute it; the latter is the best solution we can compute given a certain design heuristic.

In the incremental design approach, on the other hand, we design the new network $N_{evo}(k)$ based on the network $N_{evo}(k-1)$ from the previous environment $k - 1$. We refer to the former as the *evolved* network at environment k. The objective of the incremental design process is to identify an acceptable network $N(k) \in \mathcal{N}(k)$ that minimizes the *modification cost* $C_{mod}(N_{evo}(k - 1); N(k))$ between networks $N_{evo}(k - 1)$ and $N(k)$. For simplicity, we denote the previous modification cost as $C_{mod}(k)$.

The modification cost $C_{mod}(k)$ is defined as the cost of new design elements that are needed in $N(k)$ but are not present in $N_{evo}(k-1)$. Formally, $C_{mod}(k)$ is the cost of the design elements in the set $S_{mod}(k)$, where

$$S_{mod}(k) = N(k) \setminus [N_{evo}(k-1) \cup I(k-1)] \qquad (2)$$

abusing the notation $N(k)$ to also refer to the set of design elements in the network $N(k)$.

With the previous definitions, we can now formulate the incremental design process as:

$$C_{evo}(k) \equiv C(N_{evo}(k)), \quad N_{evo}(k) \equiv \arg \min_{N \in \mathcal{N}(k)} C_{mod}(k) \qquad (3)$$

The evolved network $N_{evo}(k)$ may not be unique in general. Ties are broken by considering a secondary objective: if two networks minimize the modification cost, select the network with the minimum total cost.[1] As in the case of optimized design, we compute the solution of the incremental design problem with a heuristic described in Section 3.

The cost of the evolved network can be expressed recursively as $(k \geq 1)$:

$$C_{evo}(k) = C_{evo}(k-1) + C_{mod}(k) \qquad (4)$$

We assume that the initial evolved network (and its cost $C_{evo}(0)$) is known. So, the cost of the evolved network at environment k has increased by the modification cost (new elements that are purchased at k).

Expanding (4), we can write the cost of the evolved network as:

$$C_{evo}(k) = C_{evo}(0) + \sum_{i=1}^{k} C_{mod}(i) \qquad (5)$$

Thus, the cost of the evolved network at environment k is the cost of the initial network plus the cost of all design elements that were purchased in the last k environments.

2.1   Metrics

We now introduce four metrics to compare a sequence of optimized and evolved networks. We also introduce two specific models of dynamic environment we consider in this paper, and quantify the rate at which the environment changes with time.

First, the *cost overhead* $v(k)$ of the evolved design $N_{evo}(k)$ relative to the corresponding optimized design $N_{opt}(k)$ at environment k is:

$$v(k) = \frac{C_{evo}(k)}{C_{opt}(k)} - 1 \geq 0 \qquad (6)$$

[1] In our computational experiments, two modification costs are rarely equal because link costs are based on distance and they are real numbers.

where the inequality is expected from the definition of $C_{opt}(k)$. [2] What is more important however is whether the cost overhead of the incremental design process gradually increases, i.e., whether the evolved networks become increasingly more expensive compared to the corresponding optimized networks. If that is the case, the incremental design process would diverge over the long-term towards extremely inefficient designs.

Second, the *evolvability* $e(k)$ is defined as:

$$e(k) = 1 - \frac{C_{mod}(k)}{C_{opt}(k)} \leq 1. \tag{7}$$

The evolvability represents the cost of modifying the evolved network from environment $k-1$ to k, relative to the cost of redesigning the network "from scratch" at time k. High evolvability, close to 1, means that it is much less expensive to modify the existing network than to re-design a new network. On the other hand, when the evolvability becomes zero or negative, it is beneficial to stop the incremental design process and design a new optimized network, i.e., clean-slate design "beats" evolution in that case.

The *topological similarity* $t(k)$ between the optimized $N_{opt}(k)$ and evolved $N_{evo}(k)$ networks is defined as the *Jaccard similarity coefficient* of the two corresponding adjacency matrices. In other words, $t(k)$ is the fraction of distinct links in either network that are present in both networks. Even though the two network design approaches we consider are different in terms of objective and design method, we are interested to know how different the resulting networks are, structure-wise, over time.

2.2   Expansion Models

We consider a specific way in which the environment changes with time: *expansion*. Specifically, the set of locations that the network has to interconnect at any time k is increasing with k. This is probably the most natural way the environment can change with time in the context of communication networks.

In the simplest form of expansion, the network size increases by only one node at each environment; we refer to this as *single-node expansion*.

We also consider a *multi-node expansion* scenario in which the network size increases once by a multiplicative factor ρ, which we refer to as *expansion factor*. Specifically, if the network size increases from n nodes to $n + m$ nodes, the expansion factor ρ is

$$\rho = \frac{n+m}{n} \geq 1. \tag{8}$$

We also compare two expansion models: *random* and *gradual*. Suppose that the set of all possible locations is $\mathcal{L}$ and that the network expands at time k to X

---

[2] The reader should note that the cost overhead metric is different than the well-known *approximation ratio*. The latter examines the worse-case solution produced by an algorithm relative to the optimal solution of a given problem. The cost overhead compares the solutions (costs) of two algorithms that aim to minimize two *different* optimization objectives.

new locations. In random expansion, the X new locations are selected randomly from $\mathcal{L}$. In gradual expansion, we select iteratively each of the X new locations from $\mathcal{L}$ so that it is the closest location to either any of the existing nodes in network $N(k-1)$ or to any of the new locations we have just added in the network.

The random and gradual expansion models represent two significantly different models in which the environment changes with time. In random expansion, the new locations can be anywhere and so it may be costly for the incremental design to adjust the previous network with only minor modifications. In gradual expansion, the environment changes in an "evolution-friendly" manner because the new locations are as close as possible to the existing network.

3   Ring Networks

In this section, we compare the optimized and incremental design approaches in the context of a ring topology. Ring networks are widely used mostly in metropolitan-area networks, as they are robust to single-node failures (two node-disjoint paths exist between any pair of nodes) and they are typically less costly than mesh networks [9]. For our purposes, the ring topology has two additional features. First, we can use an existing software package (Concorde) [8] that computes excellent approximations to the optimal ring design problem. Second, we can leverage existing analytical results to derive asymptotic expressions for the evolvability and cost overhead of incremental ring design under random and gradual expansion.

3.1   Optimized versus Incremental Ring Design

We assume that all potential nodes of the expanding ring network are located in a bounded plane region. Further, we assume that the cost of a network is equal to the sum of its link costs, and the cost of each link is proportional to its length. So, the minimum-cost ring design problem is equivalent to the NP-Hard Traveling Salesman Problem (TSP) [1]. We rely on a TSP-solving software package called Concorde [8], which is based on a branch-and-cut algorithm. Concorde's TSP solver has computed the *optimal* solutions to 106 of the 110 *TSPLIB* problem instances; the largest of them has 15,112 nodes.

Further, in the case of ring design we can use a simple asymptotic expression for the length of the optimal TSP tour. Specifically, Beardwood et al. proved that *"the length of the shortest closed path through n points in a bounded plane region of area A is almost always asymptotically proportional to $\sqrt{A \times n}$ for large n"* [3]. In our context, the length of the TSP tour is equal to the cost of the optimized ring.

In the case of random expansion, the n nodes of the ring can be anywhere in the given region and so the area A does not depend on n. Thus, the cost of the optimized ring increases as

$$C_{opt}^{rnd}(n) \sim \sqrt{A \times n} \sim \sqrt{n} \tag{9}$$

where the notation $x \sim f(n)$ means that x tends to become proportional to $f(n)$ as n increases.

In the case of gradual expansion, the area in which the n ring nodes are located increases with n. If we assume that all possible nodes are uniformly distributed with point density σ_{grd} in a bounded plane region, then the area A in which n ring nodes are located increases proportionally with n ($n = A \sigma_{grd}$). Thus,

$$C_{opt}^{grd}(n) \sim \sqrt{\sigma_{grd} \times n^2} \sim n. \tag{10}$$

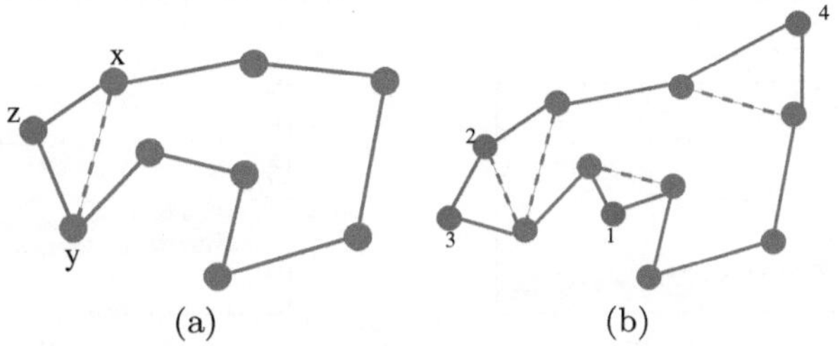

Fig. 1. Connecting new nodes to an existing ring: (a) single-node expansion, (b) multi-node expansion (we show the order in which nodes are connected)

In the incremental ring design process, we can compute the minimum modification cost under single-node expansion. Suppose that the existing ring $N_{evo}(k-1)$ has size n and we add a single extra node z at time k. The minimum modification cost will result if we connect z to two adjacent nodes x and y of $N_{evo}(k-1)$, such that

$$C_{mod}(k) = min_{(x,y) \in N_{evo}(k-1)}(||z - x|| + ||z - y||) \tag{11}$$

and then removing the edge (x, y). This process is illustrated in Figure 1(a). Note that there is no other way to connect z to $N_{evo}(k-1)$ so that the resulting network is still a ring but with lower modification cost.

In the case of multi-node expansion, we use an iterative heuristic that aims to minimize the modification cost. Suppose that the existing ring $N_{evo}(k-1)$ has size n and we add a set Z of more than one new nodes at time k. In each iteration, we select the node z from Z that minimizes the expression (11), connect z to the existing ring as we do in the case of single-node expansion, and then move z from Z to the set of nodes in $N_{evo}(k-1)$. This process is illustrated in Figure 1(b). Note that this greedy heuristic may be sub-optimal in minimizing the modification cost $C_{mod}(k)$ - an exhaustive search for the optimal solution however can be prohibitively slow.

3.2 Random Single-Node Expansion

To simplify the notation, instead of referring to an environment k, we simply refer to the ring size n. So, instead of writing $C_{mod}(k)$ we write $C_{mod}(n)$, which refers to the modification cost when the ring expands to n nodes.

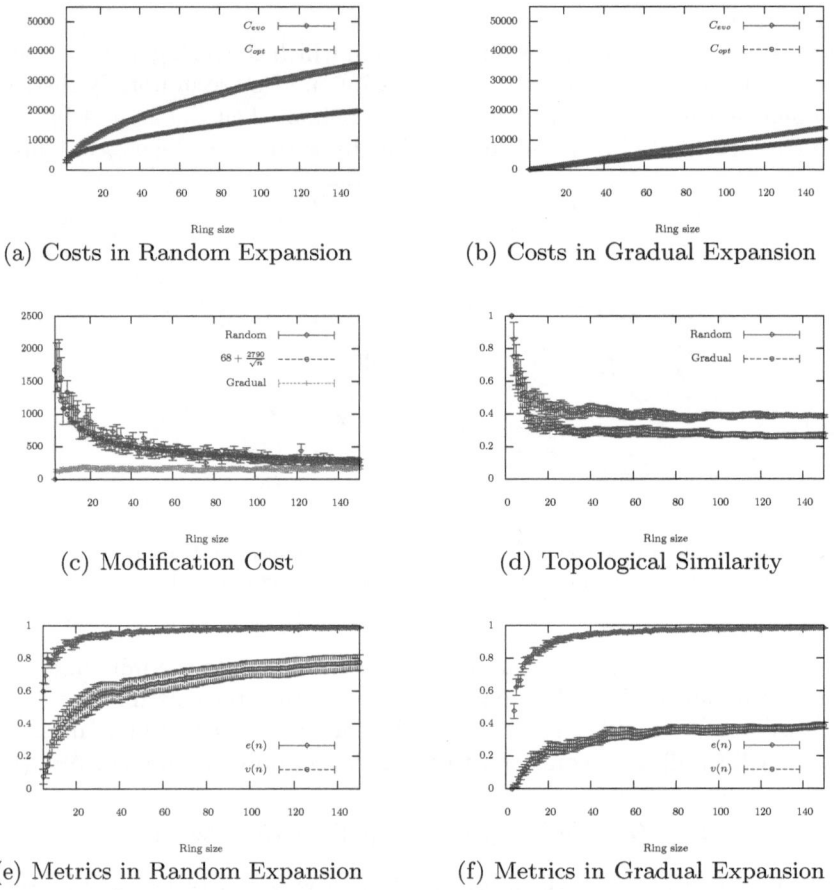

(a) Costs in Random Expansion

(b) Costs in Gradual Expansion

(c) Modification Cost

(d) Topological Similarity

(e) Metrics in Random Expansion

(f) Metrics in Gradual Expansion

Fig. 2. Results for ring network under random and gradual single-node expansion

To compute the modification cost under random expansion, we rely on a result for the nearest neighbor problem: given a set S of n points and a new point z find the closest neighbor of z in S. When n points reside in a two-dimensional region of size A with point density σ_{rnd}, the expected value of the nearest neighbor distance is $1/\sqrt{\sigma_{rnd}} = 1/\sqrt{n/A}$ [2]. The modification cost in Equation (11) can be approximated as twice the distance between the new node and the nearest node in the ring. The expected value of the latter is $1/\sqrt{n/A}$ because the n ring nodes are randomly placed in the region A. So, based on the previous approximation, the modification cost $C_{mod}(n)$ also scales as the mean nearest neighbor distance,

$$C_{mod}^{rnd}(n) \sim \frac{1}{\sqrt{n}}. \tag{12}$$

We have confirmed the validity of the previous approximation with computational results (see Figure 2(c)).

Using Equation (5), we see that the cost of the evolved network has the same scaling behavior as the cost of the optimized network,

$$C_{evo}^{rnd}(n) = \sum_{i=2}^{n} C_{mod}^{rnd}(i) \sim \sqrt{n}. \tag{13}$$

We can now derive asymptotic expressions for the evolvability and cost overhead under random expansion. From (9) and (12), it follows that

$$1 - e^{rnd}(n) \sim \frac{1}{n}. \tag{14}$$

Thus, the evolvability under single-node expansion converges to one, i.e., for large rings, the modification cost is practically zero compared to the cost of designing a new optimized ring. Also, we expect that the cost overhead will be practically constant for large values of n because both the evolved and optimized network costs scale as $\sqrt{n}$,

$$v^{rnd}(n) \sim \text{constant}. \tag{15}$$

Thus, the evolved ring does *not* become increasingly more expensive relative to the optimized ring under random single-node expansion. The exact value of the cost overhead depends on the placement of the nodes, the order in which they are added to the network, and the initial ring we start from.

We have confirmed these asymptotic expressions with computational experiments in which the optimized ring is designed using Concorde and the evolved ring is designed based on Equation 12. Figure 2(a) shows the optimized and evolved network costs (with 90% confidence intervals for the empirical results after 20 runs), Figure 2(c) shows the modification cost, while Figure 2(e) shows the evolvability and cost overhead. In the last graph, note that the cost overhead appears to increase with the ring size - we have confirmed that for larger ring sizes (larger than approximately 200 nodes) the cost overhead converges to about 0.8.

The topological similarity between the corresponding optimized and evolved rings is shown in Figure 2(d). It is interesting that the two networks share only about 30-40% of their links, when $n > 20$. Thus, the two network topologies become (and stay) significantly different during the expansion process (even though they started from the same three-node ring). Also, the topological similarity is slightly higher in random expansion compared to gradual expansion.

3.3   Gradual Single-Node Expansion

In this case, a new node is selected among all potential locations as the closest location to any existing ring node. We can rely again on the nearest neighbor problem to estimate the modification cost. Suppose that all potential new locations are uniformly distributed in the given area with density σ_{grd}. The expected value of the distance between a node in the current ring (of size n) and the nearest potential location is $1/\sqrt{\sigma_{grd}}$, which does not depend on n. The modification cost can be again approximated by twice the previous distance, and so it should not increase with n, at least for large rings,

$$C_{mod}^{grd}(n) \sim \text{constant.} \tag{16}$$

Using Equation (5), we see that the cost of the evolved network under gradual expansion scales linearly with n,

$$C_{evo}^{grd}(n) = \sum_{i=2}^{n} C_{mod}^{grd}(i) \sim n. \tag{17}$$

So, the evolvability under gradual expansion scales as in the case of random expansion

$$1 - e^{grd}(n) \sim \frac{1}{n}. \tag{18}$$

The cost overhead does not increase with n for large values of n, as in the case of random expansion,

$$v^{grd}(n) \sim \text{constant} \tag{19}$$

based on Equations (17) and (10).

The difference between random and gradual single-node expansion becomes evident in the computational results (see Figure 2). For the same ring size, the evolved and optimized costs are lower under gradual expansion, the modification cost is practically constant (and lower than under random expansion), the evolvability is not significantly different between the two expansion models, while the cost overhead is significantly lower under gradual compared to random expansion. In other words, when the ring expands in a gradual manner, we expect that the evolved network will be closer, in terms of cost, to the optimized network compared to random expansion.

3.4   Effect of Expansion Factor ρ

Suppose that we add m new nodes to a ring of size n so that the resulting network is a ring with $n + m$ nodes. The expansion factor is $\rho = \frac{n+m}{n} > 1$. What can we expect about the evolvability and cost overhead of the incremental design process as functions of ρ?

If $m \ll n$ (i.e., ρ is close to one), we can rely on the following simple approximation. Under random expansion, the modification cost $C_{mod}^{rnd}(n; m)$ when we add m new nodes will be approximately m times larger than the modification cost $C_{mod}^{rnd}(n)$ when we add a single new node at a ring of size n. So,

$$C_{mod}^{rnd}(n; m) \approx m \, C_{mod}^{rnd}(n) \sim m/\sqrt{n}. \tag{20}$$

On the other hand, the cost of an optimal ring with $n + m$ nodes will be

$$C_{opt}^{rnd}(n; m) \sim \sqrt{n + m}. \tag{21}$$

Thus, the evolvability under random expansion scales as

$$1 - e^{rnd}(n; m) = \frac{C_{mod}^{rnd}(n; m)}{C_{opt}^{rnd}(n; m)} \sim \frac{m}{\sqrt{n(n+m)}} \qquad (22)$$

that can be written as

$$1 - e^{rnd}(n; m) \sim \frac{\rho - 1}{\sqrt{\rho}}. \qquad (23)$$

As ρ increases the evolvability decreases and there is a critical expansion factor value $\hat{\rho}$ at which the evolvability becomes zero. For larger expansions than $\hat{\rho}$, it is better to abandon the existing network and redesign the ring in a clean-slate manner.

In the case of gradual expansion, the modification cost $C_{mod}^{grd}(n; m)$ will again be roughly m times larger than the modification cost $C_{mod}^{grd}(n)$ when we add only one node,

$$C_{mod}^{grd}(n; m) \approx m\, C_{mod}^{grd}(n) \sim m \times \text{constant} \qquad (24)$$

while the cost of an optimal ring with $n + m$ nodes is

$$C_{opt}^{grd}(n; m) \sim n + m. \qquad (25)$$

Thus, the evolvability under gradual expansion scales as

$$1 - e^{grd}(n; m) = \frac{C_{mod}^{grd}(n; m)}{C_{opt}^{grd}(n; m)} \sim \frac{m}{n + m} \qquad (26)$$

that can be written as

$$e^{grd}(n; m) \sim \frac{1}{\rho}. \qquad (27)$$

Again, as ρ increases the evolvability decreases, but more slowly than under random expansion. It is also important that a critical expansion factor at which the evolvability becomes zero may *not* exist under gradual expansion. Equation (27) was derived assuming that ρ is close to one, and so we cannot rely on that expression to prove that the evolvability is always positive. Computational results, however, indicate that this may be the case under gradual expansion (see Figure 3(a)).

To derive the cost overhead under multi-node expansion, we have to make two additional assumptions. First, the evolved network at size n is the same with the optimized network at size n (i.e., $C_{evo}(n) = C_{opt}(n)$). Second, because $m \ll n$, the cost of the optimized network at size n is approximately equal to the cost of the optimized network at size $n + m$ (i.e., $C_{opt}(n) \approx C_{opt}(n + m)$). Then, under both random and gradual expansion, we can write that

$$C_{evo}(n + m) = C_{evo}(n) + C_{mod}(n; m) \approx C_{opt}(n + m) + m\, C_{mod}(n) \qquad (28)$$

and so the cost overhead is

$$v(\rho) = \frac{C_{evo}(n + m)}{C_{opt}(n + m)} - 1 \approx m\, \frac{C_{mod}(n)}{C_{opt}(n)}. \qquad (29)$$

Thus, under random expansion, the cost overhead scales as

$$v^{rnd}(\rho) \sim \frac{\rho - 1}{\sqrt{\rho}} \qquad (30)$$

while under gradual expansion it scales as

$$v^{grd}(\rho) \sim 1 - \frac{1}{\rho}. \qquad (31)$$

The previous scaling expressions are derived assuming that ρ is close to one, but computational results confirm that they are quite accurate when ρ is as high as four (see Figure 3(b)).

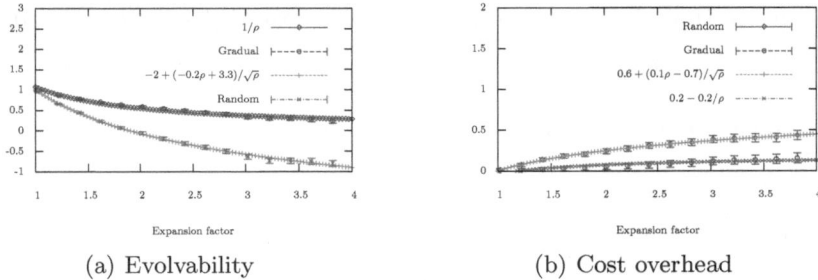

(a) Evolvability   (b) Cost overhead

Fig. 3. Effect of expansion factor under random and gradual expansion

We conducted computational experiments in which an initial ring (of size 50) is increased, under random or gradual multi-node expansion, by different factors ρ. Figure 3.4 shows the evolvability and cost overhead as functions of ρ. The previous scaling expressions, which are derived assuming that n is large and ρ is close to one, are actually in close agreement with the numerical results when ρ is as high as four. Under random expansion, the critical expansion factor is $\hat{\rho} \approx 2$, while the corresponding cost overhead is approximately 25%, meaning that even though it is then beneficial to re-design the network from scratch, the cost overhead of the evolved network is still quite low. Under gradual expansion, the evolvability does not become negative, at least in the given range of ρ, and the cost overhead is significantly lower than under random expansion.

4   Related Work

The topology design literature is extensive both in the domain of computer networks and in theoretical computer science, and it is well covered in a recent book by Pioro and Medhi [13]. The vast majority of that literature, however, focuses on optimized network design. Only few studies have focused on incremental network design (also referred to as "multi-period design"), and none of them, to the extent of our knowledge, have focused on a comparison between incremental and optimized design.

Specifically, there are some studies that focus on incremental network design [11,15,10,7,4,16,6,12,17,18] (see also section 11.2 of [13]). Those works mostly propose algorithms and optimization frameworks for incremental network design under a wide range of different constraints and objectives. Some of them consider topology design while others consider capacity expansion coupled with routing changes, and some of them consider reliability constraints while others consider limited budget constraints. None of them, however. is significantly relevant to our study because they do not compare incremental designs with the corresponding optimized designs, and they do not consider different expansion models (e.g., random versus gradual or single-node versus multi-node).

Chiang and Yang have focused on various "X-ities", such as evolvability, scalability, reliability, or adaptability in the context of computer networks [5]. They present an analytic framework to capture the notions of evolvability and scalability. They also present an Evolvable Network Design (END) Tool using dynamic programming methods to design the multi-phase deployment of a network so that early phase designs are more evolvable in later stages.

A quite different, but still relevant, study by Tero et al. [14] compared the Tokyo rail system (as an example of an optimally designed transportation network) with a natural network formed by the slime mold Physarum polycephalum. The slime mold was allowed to grow on a rectangular map of the city of Tokyo; the map contained food on the locations at which the Tokyo rail system has stations. The slime mold network grew in an incremental manner, without any centralized control or "intelligence", and it gradually covered and interconnected all food locations while refining its connectivity over time. The authors compared the two networks in terms of efficiency, fault tolerance, and cost and found that they are actually quite similar! This novel experiment implies that even a simple and incremental design process may be able to produce a network that has the cost efficiency and reliability properties that we usually only expect from optimized networks.

5 Conclusion

We can now return to the questions that were asked in the introduction and summarize our main findings. The following conclusions are supported by asymptotic scaling expressions as well as computational results for rings.

1. We formulated the incremental ring design process as an optimization problem that aims to minimize the modification cost relative to the previous network. We also identified and compared certain expansion models (random versus gradual, and single-node versus multi-node).
2. Under single-node expansion (random and gradual), even though an evolved network has higher cost than the corresponding optimized network, both costs scale similarly ($\sqrt{N}$). As a result the the cost overhead of evolved network does not increase as the network grows.

3. Under single-node (random or gradual) expansion, it is less costly to follow the incremental design approach than to re-design the network from scratch. The evolvability under basic expansion converges to one as the network grows.
4. Under multi-node and random expansion, there is a critical value $\hat{\rho}$ of the expansion factor beyond which it is less costly to abandon the existing network and re-design the network from scratch. In the gradual exapnsion, the evolvability scales as $1/\rho$.
5. Under gradual expansion, the evolvability is higher and the cost overhead is lower than under random expansion. The model of gradual expansion represents a more "evolution-friendly" dynamic environment than random expansion.

In a follow-up paper we will extend this study to capture general, mesh networks. In that case, networks are designed with additional reliability and performance constraints (e.g., a minimum number of node-disjoint paths between any pair of nodes, or a maximum propagation delay between any two nodes). We will also examine other dynamic environment models, such as iterated multi-node expansion, models in which the traffic loads and link capacities change with time as well, as well as more elaborate economic models that involve discounting or dynamic costs. It would also be interesting to see this quantitative framework and comparisons between evolved and optimized designs applied in other problems and technological domains.

References

1. Applegate, D., Bixby, R., Chvtal, V., Cook, W.: The Traveling Salesman Problem: (A Computational Study). Princeton University Press (2006)
2. Arya, S., Mount, D., Netanyahu, N., Silverman, R., Wu, A.: An optimal algorithm for approximate nearest neighbor searching fixed dimensions. J. ACM 45, 891–923 (1998)
3. Beardwood, J., Halton, J., Hammersley, J.: The shortest path through many points. Mathematical Proceedings of the Cambridge Philosophical Society (55), 299–327 (1959)
4. Carvalho, P., Ferreira, L., Lobo, F., Barruncho, L.: Optimal distribution network expansion planning under uncertainty by evolutionary decision convergence. International Journal of Electrical Power & Energy Systems 20(2), 125–129 (1998)
5. Chiang, M., Yang, M.: Towards Network X-ities From a Topological Point of View: Evolvability and Scalability. In: Proc., Allerton Conf. on Comm., Control, and Computing (2004)
6. Geary, N., Antonopoulos, A., Drakopoulos, E., O'Reilly, J.: Analysis of Optimisation Issues In Multi-Period DWDM Network Planning. In: IEEE INFOCOM (2001)
7. Gopal, S., Jain, K.: On Network Augmentation. IEEE Transactions on Reliability 35(5), 541–543 (1986)
8. Hahsler, M., Hornik, K.: TSP Infrastructure for the Traveling Salesperson Problem. IEEE/ACM Transactions on Networking 23(2) (December 2007)
9. Herzog, M., Maier, M., Reisslein, M.: Metropolitan area packet-switched WDM networks: A survey on ring systems. IEEE Communications Surveys & Tutorials 6(2), 2–20 (2004)

10. Lee, H., Dooly, D.: Heuristic algorithms for the fiber optic network expansion problem. Telecommunication Systems 7, 355–378 (1997)
11. Meyerson, A., Munagala, K., Plotkin, S.: Designing Networks Incrementally. In: IEEE FOCS (2001)
12. Pickavet, M., Demeester, P.: Long-term planning of WDM networks: A comparison between single-period and multi-period techniques. Photonic Network Communications 1(4), 331–346 (1999)
13. Pioro, M., Medhi, D.: Routing, Flow and Capacity Design in Communication and Computer Networks. The Morgan Kaufmann Series in Networking (2004)
14. Tero, A., Takagi, S., Saigusa, T., Ito, K., Bebber, D., Fricker, M., Yumiki, K., Kobayashi, R., Nakagaki, T.: Rules for Biologically Inspired Adaptive Network Design. Science 327(5964), 439–442 (2010)
15. Vajanapoom, K., Tipper, D.: Risk based incremental survivable network design (2007)
16. Wu, T., Cardwell, R., Broyden, M.: A multi-period design model for survivable network architecture selection for SONET interoffice networks. IEEE Trans. on Reliability 40, 417–427 (1991)
17. Yaged, B.: Minimum Cost Routing for Dynamic Network Models. Networks 3, 193–224 (1973)
18. Zadeh, N.: On Building Minimum Cost Communication Networks over time. Networks 4, 19–34 (1974)

A Genetic Algorithm for a Joint Routing and Scheduling Problem in Heterogeneous Networks

Hela Masri[1], Saoussen Krichen[2], and Adel Guitouni[3]

[1] LARODEC Laboratory
Institut Supérieur de Gestion de Tunis,
2000 Bardo, Tunisia
masri_hela@yahoo.fr
[2] Faculty of Law, Economics and Management,
University of Jendouba,
Avenue de l'UMA, 8189 Jendouba , Tunisia
saoussen.krichen@isg.rnu.tn
[3] Defence R and D Canada - Valcartier, Quebec (Quebec)
Peter B. Gustavson School of Business, University of Victoria,
Victoria (British-Columbia) Canada
Adel.Guitouni@drdc-rddc.gc.ca

Abstract. In this paper, we address the information routing problem in heterogeneous decision networks with known messages sizes. A decision network is a set of connected nodes in which some nodes are decision makers (DMs) requesting information, others are information providers (sources) or neutral nodes. An information might be relevant to many DMs and can be provided by different sources with different accuracies. The information value for each DM is modelled as a time dependent utility function. The problem is therefore to generate a set of efficient routing plans to satisfy the DMs' requests. The congestion problem is solved by determining the optimal transmission schedule along the chosen paths. The joint routing-scheduling problem is modelled as a bi-objective optimization problem that maximizes the overall utility and reliability of the generated paths. A multiobjective genetic algorithm (MOGA) is proposed to solve such an NP-hard problem. We show through empirical experiments that the MOGA provides a representative sample of the efficient set. We also develop an upper bound for the first objective, to validate the quality of the generated potentially efficient solutions.

Keywords: joint routing and scheduling problem, decision networks, multiobjective optimization.

1 Introduction

Information sharing in telecommunication networks is getting increasing interest especially with the great development in sensor networks, surveillance and military applications. A key problem in network management is the routing and

E. Hart et al. (Eds.): BIONETICS 2011, LNICST 103, pp. 16–28, 2012.

transmission scheduling. As these networks become larger, an efficient management of information flows is required especially for large volumes of data.

In this paper, we address the information routing problem in heterogeneous networks. A network is a set of connected nodes, where some nodes are sources or consumers (DMs) of given information, and where other nodes might just be neutral or relays of information. Information is largely exchanged in the form of messages and/or artefacts. The connectivity between nodes is characterized by a set of attributes such as the medium type, reliability, capacity and latency. Information might be relevant to many DMs, and the same information can be held by different sources. Each message has an accuracy and a size.

Assuming a centralized control, the routing problem is generally modeled as a multicommodity flow problem (MCFP), where multiple pairs of source-destination have to be managed. A linear version of this problem exists for bifurcated demand. This problem can be solved in a polynomial time [2,1,6]. However in some applications, it is required to transmit a message along a single path [3]. Such problem is called the unsplittable multicommodity problem, proven to be NP-hard [3,1]. This problem has been investigated by only few researchers in the single objective framework [4,2]. In most cases, the problem was solved using approximation algorithms [12,4]. More recently, a metaheuristic approach based on ant colony method was proposed [3]. Furthermore, adding a side constraint to the MCFP will turn the problem complexity to be NP-hard. Holmberg et al. [7] explained that in telecommunication applications there are often additional time-delay or reliability requirements on paths used for routing. These requirements may vary by communication pair, represented by different priority classes. In their paper [7], they extended the basic MCFP to include such side constraints on paths. The problem is proven to be NP-hard. As a solution approach, the column-generation method was used. All the above mentioned references mainly addressed the routing problem where the transmission rates of the source nodes are predefined (non preemptive transmission). In this case, the contention problem is handled by respecting the capacity constraint of the edges. However if we consider a centrally managed network with known data transfer sizes, the routing algorithm needs to generates transmission paths as well as a scheduling of the transmission along these paths.

Both optimal routing and scheduling problems have been largely studied in isolation. After setting the paths, the scheduling problem is solved locally. According to the message contained in the packet (how long the packet have been in the network, the distance to reach the destination, the arrival time, priority, ...), the current node will define a schedule that resolves this contention by deciding which packet to advance. The principal goal is to find a local selection rule that guarantees a bounded transmission delay. Some greedy scheduling algorithm are proven to be bounded (longest in system, shortest in system, furthest to go) independently of the network structure [5].

Recently, joint routing and scheduling problem are developed for some specific network types such as mesh networks [13] and adhoc networks [9]. The isolated use of scheduling mechanisms does not guarantee an optimal quality of service

(QoS). For this reason, the use of cross-layer optimization techniques involving joint use of scheduling and routing has been recommended in the literature [13], it yields to a better performance than the independent implementation. Link scheduling has been one of the most investigated mechanisms that takes into account the interference problem.

The review of the literature shows that multiobjective routing problems is not widely investigated. Moreover, routing is not generally coupled with the semantics of the network. The objectives considered are generally related to optimizing the cost, the shared throughput or the delay. However, and for a wide range of applications, domain specific network have usually a purpose behind information sharing.

In this paper, we study the routing problem from a new perspective. The network is considered as a decision communication network where the sub-set of nodes corresponds to different DMs. Each DM assigns a value to the requested information representing its utility. Hence, one of the QoS to optimize is the overall utility in the network. We propose a multiobjective joint routing and scheduling algorithm for selecting the paths and the time the data should start and end transmission at each link so as to optimize some performance criterion. To model the problem, a multiobjective mathematical formulation is presented, in which we maximize the overall value of information (modeled as a time dependent utility function) and maximize the reliability of the generated paths under a set of constraints (e.g., capacity, time window, compatibility, single path). Due to the complexity of the problem (NP-hard), the solution approach is a heuristic method that generates a near optimal Pareto front in a reasonable time. A multiobjective genetic algorithm (MOGA) is proposed to solve the problem. This paper presents an empirical validation of the proposed approach using a simulation environment called Inform Lab [10] that contains real instances of surveillance problem.

This paper is organized as follows. In the next section, the problem is presented. Section 3 presents the mathematical programming formulation of the joint routing and scheduling problem. Section 4 details the proposed solution approach. Section 5 provides some experimental results illustrating the efficiency of the proposed method.

2 Problem Description

We address the optimization of information routing in heterogeneous network, in the bi-objective framework. The considered problem is about sending various messages from a set of sources to different destinations. Each node in the network can be an information provider (source) or/and a DM requiring an information (destination) or simply a relay node. An edge is characterized by a capacity, a transmission delay, a type and a reliability. We assume that each DM has a soft time window to receive the requested information, and that the same information might be offered by different nodes with different accuracies and sizes. The solution to this problem consists on defining how to transmit messages from producers to consumers by assigning a source node to each request, and by generating a transmission path, such that to maximize the overall utility and reliability. In addition, a schedule is defined to ensure a load balancing over the edges of the network.

The problem can be defined by:

- **A static heterogeneous network infrastructure:** The network is composed by a finite set of connected nodes, modelled as a directed graph. A node can be an automated system like a sensor, a fusion node or a human DM. The structure of this network is fixed and the characteristics of the links are deterministic. Heterogeneity means that we might have different types of communication networks (e.g., wireless, radio, Internet). Each network type is represented by a separate set of edges. Given these statements, the network can be modeled as a directed multiple edge graph (V,E) where $V = (N_1, .., N_n)$ is the set of nodes and $E = (E_1, .., E_m)$ is the set of edges. Each arc $(i, j) \in E$ is characterized by:
  - A limited capacity c that denotes the maximum number of data units which can be transmitted from node i to node j per unit time.
  - A lead time that represents the time required to send data between the two nodes.
  - A medium type representing the compatibility of messages that can be sent along each edge.
  - A reliability
- **A set of messages:** A finite set of messages is shared by the network nodes. Each message has a compatibility constraint. Some messages can be transmitted only on a particular type of communication network (one or more). Furthermore, each DM will assign a value to each required information. This value represents the pertinence of the information for the DM, and it is modelled as a time dependent decreasing function. It is continuous, and comprised between 0 and 1, according to the time window imposed by the receiver. Figure 1 describes the form of the utility function given the time window $[a, b]$ defined by the DM. This type of function was widely used in vehicle routing problem [8]. Moreover, we assume that the utility of a DM is also dependent of the accuracy of the source node. Hence, one of the objectives is to maximize the utility of the whole network which is equal to the sum of the utility functions of all the DMs weighted by the accuracy of the chosen source for each flow.
- **A set of constraints:** A feasible solution should satisfy the following constraints:
  - Each satisfied flow (a DM asking for an information) has to be assigned to one source node.
  - A single path for each request is generated.
  - The arrival time of a message should not exceed the upper bound of the time window of the receiver.

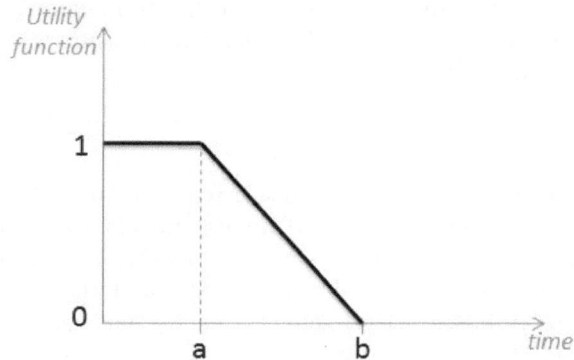

Fig. 1. The utility function of a decision maker

- Since we are managing the transmission schedule, the generated timing along the edges of a single path should be coherent.
- For a given time the total volume of data transmitted through an edge should not exceed its capacity.

Finding a solution to this problem means to provide a detailed transmission plan of different messages and their paths. A subset of requests may be satisfied according to the network resources. For each satisfied flow (i.e a pair: destination, message), we specify the assigned source and the generated path (succession of edges) used to transmit the message. A schedule of the transmission along paths allows to avoid the congestion problem. This schedule includes the transmission start time of each message on every edge.

3   Mathematical Programming Formulation

To formally model this problem, we propose a mathematical programming formulation with two objective functions. This formulation is a mixed integer non-linear program. It is an extension of the basic mathematical formulation of the MCFP (path formulation [7]). We assume that, at a given time, the central router has a perfect information about the structure of the network as well as the list of offers and requests. A set of paths relating the possible (source, destination, message) triplets are given as inputs of the model, these paths respect the compatibility constraint according to the type of message. A solution associates to each pair of (destination node, message), an assigned source node and a flow path (i.e succession of edges used to transmit the message). It also generates the schedule of transmission by specifying the transmission start time of a message along each edge of the path. Some of the requests might not be satisfied due to the time window constraint. The mathematical model simultaneously maximizes the total utility and reliability while respecting the source assignment, the capacity, time window, and scheduling constraints. Since we are dealing with multiobjective optimization, a set of non dominated solutions will be generated.

Notation

N:	the set of nodes $\{n_1..n_{|N|}\}$ of the graph
M:	the set of edges $\{m_1..m_{|M|}\}$ of the graph
I:	the set of messages $\{i_1..i_{|I|}\}$ to be shared
P:	the set of types of the networks $\{p_1..p_{|P|}\}$
acc_n:	the accuracy the node n
c_m:	the capacity of the edge m
l_m:	the lead time of the edge m
r_m:	the reliability of the edge e
$size_i$:	the size of the message i
IC:	information consumer matrix
ic_{ni}:	$=1$ if the node n is a DM requiring the message i
	$=0$ otherwise
IP:	information producer matrix
ip_{ni}:	$=1$ if the node n is a provider of the message i
	$=0$ otherwise
$U_{di}(t)$:	the utility function defined by the node d for receiving the message i at the time t
$[a_{di}, b_{di}]$:	a time window for the DM d asking for the message i
x_{di}:	$=1$ if the receiver d requiring the message i is satisfied.
	$=0$ otherwise
P^{sdi}:	the set of all paths that start at s and end at d compatible with the type of message i
p_k^{sdi}:	$=1$ if the path k is used to satisfy the flow (s, d, i)
δ_{km}^{sdi}:	$=1$ if the edge m exists on the path k used to satisfy the flow (s, d, i)
	$=0$ otherwise
T_{di}:	the total transmission time of the generated path to satisfy (d, i)
l_{sdik}:	the total number of edges composing the path of the flow (s, d, i).
$f_{sdi}^{(j)}$:	the index of the edge number j of the path of the flow (s, d, i)
ST_{sdik}^m:	the transmission's start time of the message i along the edge m in order to satisfy the flow (s, d, i)
y_{sdim}^t:	$=1$ if the edge m is used by the flow (s, d, i) at instant t
	$=0$ otherwise

The mathematical formulation is the following:

$$max\ Z_1(X) = \sum_s \sum_d \sum_i \sum_k x_{di}\ p_k^{sdi} acc_s U_{di}(T_{di}) \tag{1}$$

$$max\ Z_2(X) = \sum_s \sum_d \sum_i x_{di} \sum_k \prod_m \delta_{km}^{sdi}\ r_m \tag{2}$$

$s.t$

$$\sum_s \sum_k p_k^{sdi} \le x_{di} \qquad d = 1, .., N \quad i = 1..I \tag{3}$$

$$p_k^{sdi} \le ip_{si} \qquad d = 1, .., N \quad i = 1..I \tag{4}$$

$$p_k^{sdi} \leq ic_{di} \qquad \begin{array}{l} s = 1,..,N \quad k \in P^{sdi} \\ d = 1,..,N \quad i = 1..I \end{array} \qquad (5)$$

$$ST_{sdik}^{m} + (l_m + \tfrac{size_i}{c_m}) \leq ST_{sdik}^{m'} \qquad \begin{array}{l} s = 1,..,N \quad k \in P^{sdi} \\ s = 1...,N \quad d = 1,..,N \quad i = 1..I \\ k \in P^{sdi} \quad m = f_{sdik}^{(j)} \quad m' = f_{sdik}^{(j+1)} \\ j = 1..(l_{sdik} - 1) \end{array} \qquad (6)$$

$$T_{di} = \sum_s \sum_k p_k^{sdi} ST_{sdik}^{m} + (l_m + \tfrac{size_i}{c_m}) \qquad \begin{array}{l} d = 1,..,N \quad i = 1..I \\ m = f_{sdik}^{(l_{sdik})} \end{array} \qquad (7)$$

$$x_{di} \, T_{di} \leq b_{di} \qquad d = 1,..,N \quad i = 1..I \qquad (8)$$

$$\sum_k p_k^{sdi} \, \delta_{km}^{sdi} \, y_{sdim}^{t} = x_{di} \qquad \begin{array}{l} m = 1..M \quad d = 1,..,N \\ i = 1..I \quad s = 1,..,N \\ \lfloor ST_{fm} \rfloor \leq t \leq \lceil ST_{fm} + l_m + \tfrac{size_{f_i}}{c_m} \rceil \end{array} \qquad (9)$$

$$\sum_s \sum_d \sum_i y_{sdim}^{t} \leq 1 \qquad m = 1..M \quad t = 1..T \qquad (10)$$

$$x_{di} \in \{0,1\}, \quad p_k^{sdi} \in \{0,1\}, \quad ST_{sdik}^{m} \geq 0 \qquad \begin{array}{l} m = 1,..,M \quad d = 1,..,N \\ s = 1,..,N \quad i = 1..I \\ k \in P^{sdi} \end{array} \qquad (11)$$

The two considered objective functions are to:

- Maximize the utility of the whole network which is equal to the sum of the utility functions of each flow weighted by the accuracy of the assigned source node. The utility function is dependent of the arrival time of the required message.
- Maximize the overall reliability of the generated paths of the satisfied flows.

Constraints (1) verifies that each satisfied request, described by a pair of destination-message (d, i), is assigned exactly to one path linking its assigned source s and its destination d. Constraints (4)-(5) represent the source assignment constraints, it ensures that each a pair of (destination node, message) (d, i) should be assigned to one source node s, such that s is one of the possible producers of the required message and d a requester. Constraint (6) represents the precedence constraint of the transmission start time between the edges composing a path. Constraints (7) and (8) ensures that the arrival time T_{di} of a flow should respect the time window upper bound b_{di}. Constraint (9) states that for a given edge m and a flow (s, d, i), y_{sdim}^{t} should be equal to one for all instant t comprised between the transmission start time and the arrival time along this edge. Finally, the capacity constraint (10) means that at a given time, only one flow is passing through an edge m.

4   Solution Approach

As the performance and capacity of future communication networks will be extremely high, a much faster response is required. In addition, the complexity of the problem is NP-hard due to the single path constraint. These reasons justify the computational impracticality of exact algorithms for solving this problem. Therefore, we propose to solve it using a metaheuristic method, based on a multi-objective genetic algorithm (MOGA). This method has previously demonstrated a good ability of solving similar routing problems [3]. Starting with a random population, the individuals evolve to new solutions that approximate better the pareto front. This algorithm maintains an evolutionary population P and an external non dominated solution set P_{ND}. The basic outline of the algorithm is described, and each procedure is briefly explained.

Genetic algorithm

---

i=0
Generate first population P_0 of size N
Define the optimal schedule for each solution
repeat
$\quad i = i + 1$
Crossover: generate the set R_i of offspring of size N
Define the optimal schedule for each solution in R_i
Update the non dominated set
$P_i = P_{i-1} \cup R_i$
Evaluate and select best N solution from P_i
until (stopping criteria)

4.1   Chromosome Representation

Better efficiency of GA-based search could be achieved by well defining the chromosome representation and its related operators so as to generate feasible solutions and avoid repair mechanism. Figure 2 depicts the structure of a chromosome. The adopted representation is a variable length with bounded solution size; since the size of a substring vary with the number of the edges figuring in the path. A chromosome is coded as a vector of R substrings where R is the number of requests. Each substring is composed of two parts::

- A bit describing if this request is satisfied or not.
- Two vectors describing the combination of path assignment and transmission scheduling decisions, expressed by the list of indexes of the used edges composing the path and the sequence of the transmission start time on these edges.

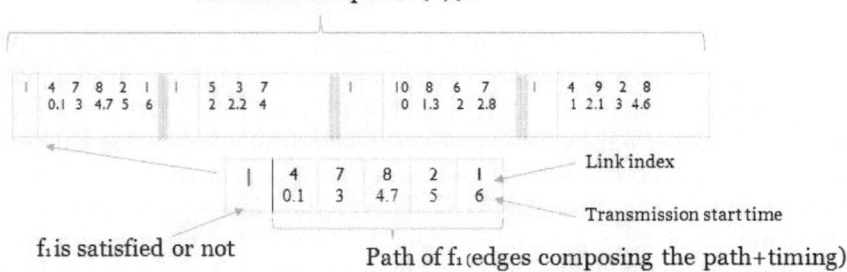

Fig. 2. Chromosome representation

4.2 Generating the First Population

Population initialization is a crucial task in evolutionary algorithms because it can affect the convergence speed and the quality of the final solution [11]. In order to find the initial population, a greedy algorithm based on nearest neighbour method is used, weights are randomly generated in order to aggregate the two objective functions. After setting the size of the population, for each solution, we generate separately a feasible path for each request. The paths are dynamically built such that a partial solution is constructed sequentially following a probabilistic model. The choice of a neighbour depends on the reliability and utility of an edge. These two measures are aggregated according to random weights. A random wheel selection is applied to choose the best edge (i.e the next node to be explore in the path). We use a reverse path construction strategy so that we start from the destinations until reaching an adequate source. Then, for each generated chromosome of the first population, a greedy scheduling algorithm is applied to find the transmission start time of the messages on each edge.

4.3 Crossover

In general, the crossover operator is regarded as a main genetic operator ensuring the diversification mechanism, and the performance of the genetic algorithms depends, to a great extent, on the performance of the crossover operator used. Conceptually, the crossover operates on two chromosomes at a time and generates offspring by combining features of both chromosomes. The proposed algorithm applies the random key method for the crossover, the offsprings are produced from two parent solutions following these steps.

1) two chromosomes ch_1 and ch_2 are chosen from the current population P_i
2) a probability p is randomly generated
3) if $p < p_c$ then the substring for a new child is chosen from the first chromosome $ch1$ otherwise it is taken from the second chromosome ch_2
4) Repeat 2 and 3 until reaching the last substring.

Only the paths and the two first parts of the substring are involved in the crossover. Therefore, the schedule will be modified according to the result of the crossover.

4.4 Generating the Schedule

After generating N new offsprings, a schedule is defined for each new solution. The schedule is based on a greedy algorithm. Given a feasible solution where the paths are already set, the algorithm computes for each request the arrival time of the corresponding message. Then, it accordingly ranks the flows to be satisfied first, by subserving the requests that have the most restrictive time windows. After defining the ranking, the requests are satisfied in order of their rank. Therefore, the edges are considered as resources to be shared. A list recording the intervals of occupancy of each edge is managed. If the arrival time at a destination exceeds the time window, the corresponding request is considered as not satisfied.

4.5 Selection Procedure

After generating the offsprings, a non domination sorting is applied in order to evaluate the obtained solutions. The N best solutions are selected to be the population of the next iteration. This procedure ranks the solutions of the current population based on non-domination. All the non-dominated individuals in the current population are placed at the top of a list and assigned rank 1. These solutions are removed from the remaining population and the next set of non-dominated solutions is identified and assigned rank 2. The process is repeated until the entire population is ranked. The top N individuals in the list are then selected for the next generation.

5 Experimental Results

In order to evaluate the proposed algorithm, a computational analysis is performed in a test bed using real instances of surveillance problems. As a simulation environment, we used INFORM-Lab Simulation testbed [10]. This environment enables to execute different algorithms for distributed information fusion and dynamic resource management. It contains different surveillance vignettes with different scenarios. The proposed algorithm is coded in java. The experiments were conducted on a core 2 1.5 GHZ laptop. The instances are generated based on a real network: the US National Science Foundation (NSF) [3], containing 14 nodes and 42 directed arcs.

The simulation parameter to vary is the number of messages to be shared. For each message we generate a random number of requests and offers. For each request we assume that the time window is assigned randomly. Also for each offer, we generate randomly the size of the information as well as the accuracy of the information (random but positively correlated to the size). A total number

of 10 instances are generated, by varying the number of messages between 10
and 100. Table 1 describes the configuration used to generate each instance. We
assume that each parameter is uniformly distributed in the defined interval.

Table 1. Experiments design

Parameter	Interval
Number of requests	$[1..2]$
Number of offers	$[2..3]$
Message size	$[1..10]$
Accuracy	$[0.5..1].(1 - \frac{1}{size})$
Time window	$a \in [0, 50]$
	$b \in [50, 300]$

We choose the maximum number of iterations as the termination criteria, set
to 1000 iterations. Also if there is no improvement after 100 iterations, the algorithm stops (i.e if all the generated solutions over the last 100 iterations are dominated). The crossover probability is set to 0.5, and the population size is equal
to 100. Each Instance is solved with 10 independent runs. we report in table 2
the average of the following measures: number of messages to be shared $NbMsg$,
number of offers $NbOff$ and requests $NbReq$, the CPU time, the size of the non
dominated set $|P_{ND}|$, the best utility U and reliability R of all the generated non
dominated solutions, the CPU time for generating the first population FP

Table 2. Experimental results

	Problem description			Results					Upper Bound	
Problem	Nb Msg	Nb Off	Nb Req	CPU(s)	$|P_{ND}|$	U	R	FP CPU(s)	UB_U	Gap_U
1	10	24	11	0.43	7	8.65	3.25	0.23	10.29	0.19
2	20	53	28	0.75	12	19.91	7.14	0.39	24.88	0.25
3	30	74	32	1.23	9	31.5	9.65	0.61	39.67	0.26
4	40	115	52	1.74	14	41.5	15.89	32.4	53.53	0.29
5	50	130	86	3.336	17	48.22	22.7	1.38	66.54	0.38
6	60	134	96	4.69	13	48.8	24.38	1.96	67.34	0.45
7	70	174	102	6.09	21	52.08	27.42	2.48	79.16	0.52
8	80	196	135	6.69	24	67.72	29.6	2.69	121.21	0.79
9	90	238	142	7.03	23	66.87	30.88	2.72	117.2	0.75
10	100	264	153	9.08	26	77.61	32.43	3.6	144.35	0.86

Furthermore, we propose an upper bound for the first objective function dealing with the total utility UB_U by relaxing the constraint of capacity. In this case
the contention problem is discarded and the optimal solution will corresponds
to the sum of utilities of the shortest paths for each request (e.g the shortest
path has the lowest delay so that it guarantees the best utility). Hence, we solve

separately for each request the shortest path problem by trying all the possible combinations with the different sources offering the required message. For each request, the network is reduced by deleting the edges that are not compatible with the message type, then the Dijkstra algorithm is used to find the shortest path between the requester node and all the possible sources. The best path is chosen. Hence, UB_U will correspond to the sum of the utilities of the different generated paths.

$$UB_U = \sum_d \sum_i Utility(min_{s \in N | ic_{si}=1} \; ShortestPath(d,i,s))$$

We report, in table 2, the gap representing the average deviation of the solutions generated by the MOGA algorithm from the upper bound values.

By analyzing these results, we can notice that:

- The CPU time seems to grow polynomially with the problem size. We also notice that the CPU time for generating the first population represents approximately 40% of the total CPU time required to solve each instance. This fact is due to the constructive nature of the procedure used in the first population.
- Generally, when the number of messages increases, the number of possible paths' combinations grows. Hence, the number of diversified potentially efficient solutions becomes larger. For example, if the number of messages is equal to 10, the average size of the non dominated set is 7. However, if 100 messages have to be routed the algorithm generates 26 potentially efficient solutions. Nevertheless, the size of the non dominated set seems to be limited for all the instances. This fact shows that the two considered objectives are conflicting but not so divergent. Because the paths with the lowest number of hops would likely have the best utility and reliability.
- By analyzing the values of the utility's upper bounds and the best utilities generated by MOGA algorithm for the 10 instances, we notice that for the small sized instances the average gap is about 0.25 which is reasonable. However, if the number of messages to be routed increases, the network resource sharing will be intensified. Therefore, using the shortest paths as un upper bound will give an unrealistic estimation of the overall utility value. The less contention we have in the network, the tighter is the proposed upper bound. This explains the increase of the gap values (Gap_U) for the instances 6,..,10 of table 2.

6   Conclusion

In this paper, we propose to model the joint routing-scheduling problem in a decision communication network with known data transfer sizes. We proposed a bi-objective mathematical formulation modeling a preemptive routing. Two main objectives have to be optimized: maximize the overall information value and the reliability of the paths, under some constraints. Compared to the existing literature, the proposed approach seems to be very promising since we are handling the routing problem from a different perspective by including the value of information as a principal metric. To cope with the complexity of the problem (NP-hard), we proposed an adaptation of a genetic algorithm. The algorithm efficiency was validated by computational experiments on different instances. A upper bound approximating the total utility was also proposed. The gap between the generated solutions and the upper bound is considerably reasonable. This fact shows that the obtained set of potentially efficient solutions gives a good approximation of the Pareto front. As a future work, more tests over other network topologies will be performed.

References

1. Baier, G., Köhler, E., Skutella, M.: On the k-Splittable Flow Problem. In: Möhring, R.H., Raman, R. (eds.) ESA 2002. LNCS, vol. 2461, pp. 101–113. Springer, Heidelberg (2002)
2. Barnhart, C., Hane, C.A., Vance, P.H.: Using branch-and-price-and-cut to solve origin-destination integer multicommodity flow problems. Oper. Res. 48(2), 318–326 (2000)
3. Barnhart, C., Hane, C.A., Vance, P.H.: An ant colony optimization metaheuristic for single-path multicommodity network flow problems. Journal of the Operational Research Society 61(9), 1340–1355 (2010)
4. Bley, A.: Approximability of unsplittable shortest path routing problems. Networks 54(1), 23–46 (2009)
5. Bordin, A., Kleinberg, J., Raghavan, P., Sudan, M., Willamson, D.P.: Adversarial queuing theory. J. ACM 48(1), 13–38 (2001)
6. Freng, G., Douligers, C.: A neural network method for minimum delay routing in packet switched networks. Computer Communications 24, 933–941 (2001)
7. Holmberg, K., Yuan, D.: A Multicommodity Network-Flow Problem with Side Constraints on Paths Solved by Column Generation. INFORMS Journal on Computing 15(1), 42–57 (2003)
8. Ioannou, G., Kritikos, M., Prastacos, G.: A problem generator-solver heuristic for vehicle routing with soft time windows. Omega 31(1), 41–53 (2003)
9. Lu, G., Krishnamachari, B.: Minimum latency joint scheduling and routing in wireless sensor networks. Ad Hoc Networks 5(6), 832–843 (2007)
10. MacDonald, Dettwiler and Associates Ltd, Inform Lab Wiki, https://xwiki.mdacorporation.com/InformLab/www/wiki/view/Main/WebHome
11. Shahryar, R., Hamid, R.T., Magdy, M.A.S.: A novel population initialization method for accelerating evolutionary algorithms. Computers and Mathematics with Application 53, 1605–1614 (2007)
12. Skutella, M.: Approximating the single source unsplittable min-cost flow problem. In: Proceedings of the 41st Annual Symposium on Foundations of Computer Science, p. 136 (2000)
13. Wang, X., Garcia-Luna-Aceves, J.J.: Distributed joint channel assignment, routing and scheduling for wireless mesh networks. Computer Communications 31(7), 1436–1446 (2008)

The Protein Processor Associative Memory on a Robotic Hand-Eye Coordination Task

Omer Qadir, Jon Timmis, Gianluca Tempesti, and Andy Tyrrell

Intelligent Systems Group, Department of Electronics, University of York,
Heslington,York YO10 5DD, UK
oq500@ohm.york.ac.uk, jtimmis@cs.york.ac.uk
http://www-users.york.ac.uk/~oq500/

Abstract. The PPAM is a hardware architecture for a robust, bidirectional and scalable hetero-associative memory. It is fundamentally
different from the traditional processing methods which use arithmetic
operations and consequently ALUs. In this paper, we present the results
of applying the PPAM to a real-world robotics hand-eye coordination
task. A comparison is performed with a nearest neighbour technique
that was originally used to associate the same dataset. The number of
memory load/store operations and the number of ALU operations for the
nearest neighbour algorithm is compared with the corresponding PPAM
which acheives the same association. It was determined that 29 conflict
resolving nodes were required to fully store and recall the entire dataset
and the maximum number of memory locations required in any node was
160, with the average and quartiles being much lower.

Keywords: Self Organising, PPAM, Associative Memory, Protein
Processing, hetero-associative, SABRE, robotics, catcher-in-the-rye.

1 Introduction

This work stems from research into parallel architectures targeted towards problems in the domain of Artificial Intelligence (AI). Although, many architectures
exist that attempt solutions for problems in this domain (Amaral and Ghosh,
1994; Anzellotti et al., 1995; Lee et al., 2007, 2006; Teich, 2008; Toffoli, 1984),
these are based on Arithmetic and Logic Units (ALUs) and often are simply Von
Neumann style processors in parallel, and suffer from the well known Von Neumann bottleneck. An AI application may be defined as one that tries to achieve
behaviour that will be displayed by intelligent beings like humans (Turing, 1950).
    Some of the desired features of such applications are adaptability (in the pres-
ence of new stimuli), robustness (in the presence of errors), real-time results and
scalability. In the past, AI was seen as being limited by the speed of the hardware (Moravec, 1998). This assumption is being challenged more and more often
today, particularly since it has been shown that signals in the human nervous system travel in the KHz range (as opposed to the GHz range in computers today).
In any case, most will agree that the hallmark of an AI application is complexity.

E. Hart et al. (Eds.): BIONETICS 2011, LNICST 103, pp. 29–43, 2012.

The traditional computation paradigm is to implement simple (mathematical) operations in ALUs, which are accessed by instructions to a control unit. This is further accessed through (potentially multiple) software/firmware wrapper layers, which may be an Operating System or a Virtual Machine or both. Therefore, implementing complex AI applications on traditional machines not only involves mapping an individual complex operation to multiple simple mathematical operations, but also each operation passes through multiple software layers to reach a hardware unit (ALU) that is accessed through yet another hardware wrapper (the control unit). Traversing all these layers is a severe handicap for an application that intends to operate in real-time. In Qadir et al. (2010) we proposed a novel architecture called the Protein Processor Associative Memory (PPAM), that uses hetero-associative recall to achieve artificial intelligence and explores the effect of moving computation into memory. The objective was to determine whether any meaningful computation could be achieved purely through memory operations and without performing any arithmetic operations. Qadir et al. (2011a) compares the PPAM with the original BAM by Kosko (1988) and also its popular, optimal variant called PRLAB (Oh and Kothari, 1994). Although the PPAM successfully navigated a robot in an environment with obstacles and proved to be better in many respects than the BAM and PRLAB (Qadir et al., 2011a), it still had issues for a hardware implementation. As the entire argument is based on moving computation away from arithmetic operators, hardware implementability is essential since implementing a software version of the PPAM would continue to use the ALU and void the entire effort. Some changes were required before the PPAM could be implemented in hardware and Qadir et al. (2011b) presents this modified version along with details about the hardware architecture.

In this paper, we test the performance of the PPAM on a hand-eye coordination task for a robot, originally performed as part of the *catcher-in-the-rye* project (Hülse et al., 2009; Hülse et al., 2010). A robotic arm places objects in (and retrieves them from) its *reach space*, while a stereo vision system tries to identify the location of the object in its *vision space*. The task is to associate the vision space with the reach space so that when one is known, the other can be *recalled*. The hardware platform described in Hülse et al. (2009); Hülse et al. (2010) was used at the University of Aberystwyth for the same task and is the source of the dataset used in this paper. We present the results of the performance of the PPAM on this dataset and compare the number and type of operations required to achieve similar results. We end with some thoughts on the larger implications of this on the hardware resources required for such a task.

The rest of this paper is structured as follows. Section 2 outlines the principles of protein processing and provides a short description of the hardware as discussed in previous publications. Section 3 describes the experiment setup as well as the strategy used by Hülse et al. (2010). Section 4 presents the results observed from using the PPAM while section 5 compares them with those presented by Hülse et al. (2010) and also discusses the differences in the architectures and its implications on the hardware. Section 6 summarises, concludes and discusses future directions.

2   Protein Processing Associative Memory

A detailed description of the biological inspiration, learning mechanism, connection topology, and conflict resolution strategy for the PPAM may be found in Qadir et al. (2011a,b). A brief summary is presented here.

The PPAM takes inspiration from the biological Genetic Regulatory Network (GRN) and also the biological Neural Network. It is composed of a large number of simple nodes, each of which receive input from the environment (data from one of two variables to be associated) and also from N neighbouring nodes (data from the second variable after processing by neighbouring nodes). Nodes use the Hebbian principle (Hebb, 1949) to determine if two pieces of data are related and uses a GRN inspired model to decide how (and where) to store the data in the nodes so that it can be recalled quickly and correctly, even in the presence of faults.

2.1   Network Topology and Structure

A sample dataset composed of abstract symbols encoding some arbitrary dataset is shown in table 1 (Qadir et al., 2011b) where 0x represents hexadecimal numbers. In order to store this dataset consider a network of 4 abstract nodes as shown in figure 1a. An individual node in Figure 1a (for example X) has connections to D other nodes, where D is the number of dimension of the other variable (which is 2 in this case). Each of these abstract nodes in figure 1a (for example X) can be further expanded into multiple other nodes as shown in figure 1b. The horizontal axis has D number of nodes and the vertical axis has C number of nodes where C is the maximum number of conflicts that are being catered for. Conflicts occur if nodes observe the same neighbourhood inputs in conjunction with different output values from the node itself. This can be a result of input variables not being related, or the relationship being one-to-many (or many-to-many or many-to-one) rather than one-to-one and so on.

All nodes in figure 1b receive the same external (environmental) input – and therefore fire with the same output value when an external input is applied. Each node receives only one neighbour-node input – where neighbour-node is a node from the other variable which is being associated with this variable.

Table 1. Cartesian Polar Encoded Dataset

X	Y	R	θ
1	5	0x4	0xA
1	6	0x4	0x9
2	5	0x4	0xC
2	6	0x3	0xA
3	5	0x3	0xC
3	6	0x2	0xB
4	5	0x2	0xC
4	6	0x2	0xC

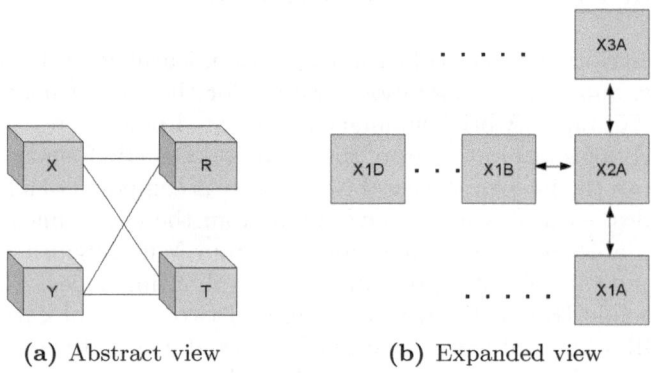

(a) Abstract view   (b) Expanded view

Fig. 1. View of the PPAM nodes (Qadir et al., 2011b)

In addition, each node in the same column receives the same neighbour-node input. Therefore, with reference to the sample dataset, there would be 2 columns in figure 1b, since there are 2 dimensions in the polar coordinate variable (R, θ). Also, the nodes in column A would receive input from nodes in one dimension of the polar coordinate variable (for instance R) while the nodes in column B would receive input from nodes in the other dimension (for example θ). Note that each node observes input from only one neighbour-node and so, irrespective of the number of dimensions of the variables, it requires only one symbol wide memory. Therefore, since the structure of a node and particularly its memory width is independent of the dimensions of data, this allows the architecture to be scalable.

2.2   Conflict Resolution

Conflicts occur when a node tries to associate different neighbour-node inputs with the same external input or vice versa (same neighbour-node input with different external inputs). The first is called a conflict-in-input, while the latter is a conflict-in-output. Although conflicts will certainly occur when there is a many-to-one (or one-to-many or many-to-many) relationship in the dataset, conflicts may also occur in a dataset where there is only a one-to-one relationship. Consider the first two tuples in table 1. Although (X, Y) together are unique, a single X node observing θ values would observe first A and then 9 while its external input would stay 1. Thus it would observe conflicting values. If there are C number of conflict resolving nodes, then C conflicts in the dataset can be resolved.

Nodes *differentiate* (inspired from GRNs) to resolve conflicts. Upon observing a conflict, some conflict-resolving nodes (in one column in Figure 1b) update to the new (conflict generating) tuple, while other nodes maintain the old (previous) tuple. During recall, nodes perform a custom lookup operation inspired from the protein transcription process in a biological cell. The output of nodes is reinforced by other nodes generating the same output so that a confidence level can be estimated for a recalled value. The memory in each node can be likened to the DNA in a cell and this self-modification of memory can be likened to changing genetic expression, which is the basis of genetic differentiation. Nodes are updated in a cascaded fashion. All nodes in one row may be updated simultaneously or in any sequence as long as they are all updated before the nodes in other rows start or after the nodes in other rows end updating. Therefore, memory writes are synchronous, however memory recalls are asynchronous and occur as soon as input is applied. For more details of the architecture and implementation, the reader is referred to Qadir et al. (2011b).

3 Experiment Setup

Figure 2 illustrates the setup used to perform the experiment in the original Catcher-in-the-rye project (Hülse et al., 2009; Hülse et al., 2010). As mentioned in Section 1, the task is to associate the vision space of the camera with the reach space of the robotic arm. Thus, when one is known, the other can be *recalled*. The vision system is mounted on a pan–tilt–verge unit and is composed of 2 cameras, each providing 1032×778 RGB images at the rate of 25 fps. Hülse et al. (2010) reduced the images to 129×62 pixels by taking the mean of 8×8 pixel blocks. Object detection was performed by filtering the image for blue colour, and tilt–verge movements were performed in the camera to saccade to the detected object. Since the pan movement for the camera is not used, the vision system has 3 degrees of freedom (one verge movement for each of 2 cameras and one tilt movement for both cameras together) and thus 3 dimensions in the variable. Although the robotic arm has 7 degrees of freedom, the location of an object in the *reach space* of the arm can be represented using polar coordinates on a 2-dimensional plane. Thus the second variable has 2 dimensions. Therefore, each entry in the dataset is composed of two variables; the first is a

Fig. 2. Robotic arm and camera setup (Hülse et al., 2010)

3-dimensional variable representing motor control for saccading the vision system and the second is a 2-dimensional variable representing polar coordinates for the robotic arm reach location.

3.1   Original Catcher-in-the-rye

The approach used by Hülse et al. (2010) is to store selected associative pairs from the dataset and use Euclidean distance to recall the closest matching pair. Each associative pair is tested to see if it should be added to the *map* and if so a new *link* is created. Old links that haven't been refreshed in a specified number of cycles are removed from the map. The algorithm is summarised in Algorithm 1.

The location of objects in the reach space is represented by polar coordinates (d, α), as shown in Figure 2, that range as follows: $30 \leq d \leq 60$ cm; $-1.4 \leq \alpha \leq 1.4$ radians.

$$D = \sqrt{[d_1 \cdot sin\alpha_1 - d_2 \cdot sin\alpha_2]^2 + [d_1 \cdot cos\alpha_1 - d_2 \cdot cos\alpha_2]^2} \qquad (1)$$

The distance D between points in the reach space in Algorithm 1 can be measured according to equation 1, which can be seen to include several mathematical operations.

Benchmarks to evaluate the computational cost of trigonometric functions vary greatly, depending upon the hardware architecture and the method used to compute the function. Intel architectures are used as the standard here, since they provide hardware support for calculating trigonometric functions using the native instructions (i.e. *fsin, fcos* and *fsincos*). Despite the hardware support however, Intel (2011) shows that clock latency for sine is 160–200 cycles with a throughput of 130 cycles; the latency for cosine is 180–280 cycles with a throughput of 130 cycles; and the latency for sine and cosine together (of the same angle) is 170–250 cycles with a throughput of 140 cycles. Here Intel (2011) defines clock latency as the number of cycles to complete execution of the instruction and throughput as the number of cycles to wait before the instruction can be issued again. The exact number of cycles is dependent upon the range of the input data and the processor model. This estimate does not include the overhead from the branch instructions or the load/store instructions required to manipulate the data. Furthermore, the calculations are performed on floating point numbers, thus requiring a Floating Point Unit. Hülse et al. (2010) uses a training set of 300 elements, and shows that setting $Q = 300$ and $T = 0.0$ produces the best results, since these values result in the map storing all the associative pairs in the training set. Thus, for an arbitrary associative pair, at most N (300) iterations of the loop (with trigonometric calculations) may need to be executed to find the closest matching link. A more detailed comparison of the number and type of instructions is presented in Section 5.

Algorithm 1. Inferred pseudo-code for Catcher in the Rye

```
// T is Tolerance for acceptable error in mapping
// Q is age threshold for old links
// M is map of links -- empty initially
// V is the 3 dimensional variable representing the vision system movement
// R is the 2 dimensional variable representing the Robotic arm movement
// N is the total number of elements in the training set
for each of N elements in the training set do
    Arm places object at a known position;
    Vision saccades to focus on the object;
    for i ← 0 to N do
        // Calculate Euclidean distance between V and M[i]
        D ← V − M[i];
        if D <= T then
            break loop as link found;

        if age of M[i] > Q then
            remove M[i] from M;

    if link found then
        zero age of link;
    else
        add a new link {V, R} to M;
```

3.2   PPAM

The 300 element dataset used by the catcher-in-the-rye experiments described in Section 3.1, was used as input to the associative memory. The dataset was composed of 3-dimensional floating point numbers for camera movement and 2-dimensional floating point numbers for arm movement. Tables 2 and 3 present the relevant features of the dataset for the camera movement variable and the robot arm movement respectively.The data captured as part of the original Catcher-in-the-rye experiment had been rounded to varying precision by by Hülse et al. (2010). Camera movement data was rounded to 3 decimal places and the angle α in robotic arm movement was rounded to 2 decimal places. The distance d did not need to be rounded. The data was encoded into abstract symbols as follows:

Table 2. Camera movement dataset features

Property	tilt	verge left	verge right
Min	-0.486	-0.334	-0.524
Max	-0.322	0.436	0.247
Range	0.164	0.770	0.771
number of discrete values	164	770	771

Table 3. Robot arm movement dataset features

Property	distance (d)	angle (α)
Min	33.0	-01.40
Max	58.0	01.40
Range	25.0	02.80
number of discrete values	25	280

- Each point in the data was normalised by subtracting the corresponding minimum value and then converted to abstract integer symbols.
- All the data was multiplied by 1000 to convert it to integer numbers (each of which represents an abstract symbol).

At first glance, this might seem like the PPAM is simply off-loading the computation (for encoding from real numbers to integers) to a companion controller, and claiming freedom from floating point units because they are implemented elsewhere in the system. However, an analysis of the system as a whole indicates that this is not the case. Analogue inputs arrive from the physical world and need to be encoded into digital data using some representation (typically 32-bit floating point) before they can be used by the system. Even when using integer representation for inputs, it is still necessary to perform floating point arithmetic because often division is involved. Note that regardless of the representation used, *some* precision will be lost through the process of quantisation, therefore from this perspective alone, both representations are equally valid. Although the process described above of encoding the dataset into abstract symbols appears to be re-encoding the dataset, in a complete PPAM system, the external inputs would not need to be represented as floating point numbers first and could directly be encoded as integers.

Experiments were performed with an increasing number of conflict resolving nodes to determine the number of nodes required to correctly store and recall the complete dataset. Data was collected to profile memory usage for the PPAM and results are presented in Section 4.

4   Results

In order to determine the number of nodes required to accurately store and recall the 300 element dataset, experiments were performed using node configurations with 10, 20, 21, 22, 23, 24, 25, 26, 27, 28, 29, 30, 50, 100 and 200 conflict resolving nodes. This set was chosen empirically while searching for the value of C (number of Conflict-Resolving nodes) that would accurately recall the entire dataset. Figure 3 plots the number of data points that were correctly stored and recalled for the different node configurations.

It can be seen that adding more conflict resolving nodes initially makes a large impact on the memory capacity but provides diminishing returns. Figure 4b shows the total number of memory locations available and the number actually used in the PPAM on a logarithmic scale. This "total number of locations used" is the sum of all memory locations used in each node. The distribution for the number of memory locations used in each node can be seen in Figure 5, which shows that the maximum number of memory locations used in *any* node for *any* configuration was 160 whereas the upper quartile is half that and the median is even lower. Furthermore, two unexpected results can be seen from the analysis of the memory usage. First, the ratio of "memory locations used", to "the total number of memory locations" in the PPAM is constant irrespective of

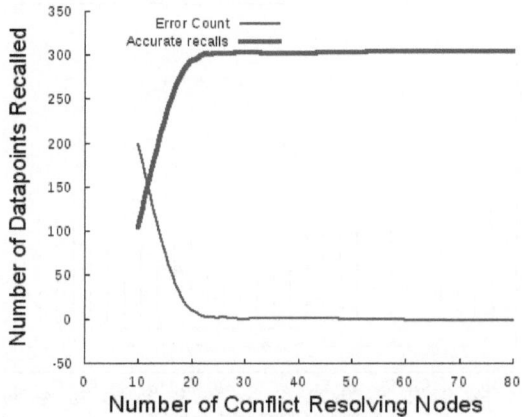

Fig. 3. Performance of PPAM as C is varied ($C = 29$ is best case)

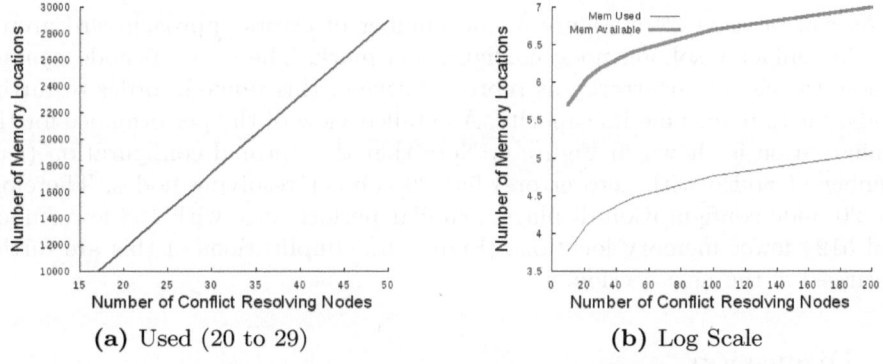

(a) Used (20 to 29)   (b) Log Scale

Fig. 4. Memory profiling

the size of the PPAM (Figure 4b). Second, although the structure of the PPAM dictates that there is a linear relation between the number of nodes and the amount of memory in the PPAM, Figure 4a shows that there is also a linear relationship between the number of memory locations actually used and the size of the network. The reason for this can be seen from Figure 5 which shows that the distribution of memory utilization per node does not change significantly with respect to the number of conflict resolving nodes used. Thus, increasing the number of nodes seems to have a linear effect on the memory utilisation but an exponential effect on the performance (Figure 3). This means that increasing the number of nodes initially has a small cost in terms of memory and a large advantage in terms of performance; whereas the opposite is true in order to learn the last few elements of a dataset. Implications of this are discussed further in Section 5.

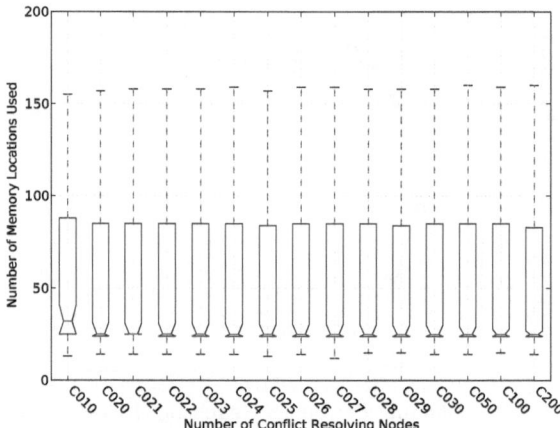

Fig. 5. Memory locations used per node as C is varied

As can be seen from Figure 3, the number of errors approach zero around the 20 conflict resolving node configuration mark. The $C = 20$ node configuration was chosen arbitrarily as representative of this range in order to further analyse it to determine its capacity. A detailed view of the performance for this configuration is shown in Figure 6. Note that the optimal configuration (least number of nodes with zero errors) has 29 conflict resolving nodes. Therefore, the 20 node configuration displays a similar performance with 108 fewer nodes and 5194 fewer memory locations (Figure 4a). Implications of this are further discussed in the next section.

5   Discussion

Section 4 shows that the PPAM is capable of storing and correctly recalling the entire dataset. Table 4 compares the number and types of operations used in the PPAM and the more traditional method used by the catcher-in-the-rye. It does not consider the branch operations implicit in any software implementation of the algorithm, since these would be removed if the algorithm was implemented in hardware. The catcher-in-the-rye implementation is assumed to be optimized such that calculations are not repeated and results are considered to be stored in temporary memory (CPU registers). Memory access (read/write) does not include access to these temporary locations or to any other *working* memory. Furthermore, memory requirements (or operations) for stimulating nodes or providing input is not included. In addition, the multiplication and addition operations for calculating indexes for the values in the link-map for the catcher-in-the-rye are also not included in Table 4 because this is considered to be similar to the significantly simpler and smaller number of memory index counting operations for the PPAM. As mentioned in Section 3, assuming there are 300 elements in the map for the catcher-in-the-rye ($Q = 300$, $T = 0.0$), it can be seen from algorithm 1 that a

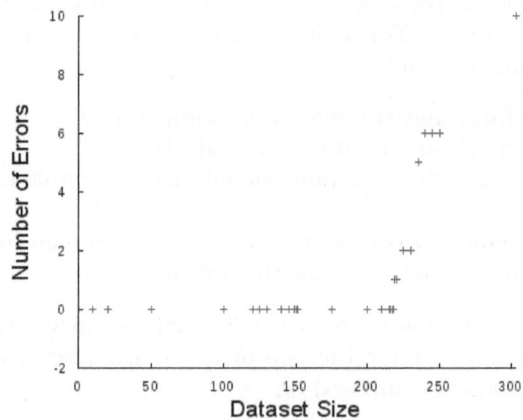

Fig. 6. Error count for $C = 20$ node PPAM configuration

single memory recall operation will take 300 iterations in the worst case and 1 iteration in the best case. Assuming an evenly distributed random distribution for the memory recall operations, the average can be assumed to be 150 iterations. Each iteration requires a number of floating point operations, which depend on whether the robot arm movement is being recalled or the vision movement is being recalled. Since the framework used in Hülse et al. (2010) looks up robot arm movements using the vision system as input, the same is used for the purpose of this comparison. Note that this assumption is not required for the PPAM since the PPAM treats every variable in exactly the same way. During memory recall, each iteration of the loop in algorithm 1 includes:

- 1 Euclidean distance measurement operation (for D)
- 1 threshold comparison operation (T)
- 3 memory read operations (3 dimensions of the vision variable in $M[i]$)

Table 4. Comparison of Catcher-in-the-rye and PPAM

	Catcher-in-the-rye	PPAM
Trigonometric	up to $S \times 4$	0
Multiplications	up to $S \times 6$	0
Square root	up to S	0
Additions	up to S	0
Subtractions	up to $S \times 2$	0
> Comparisons	0 * or up to S †	0
<= Comparisons	up to S	0
== Comparisons	0	up to $S \times H$
Memory reads	up to $S \times 3$ * or up to $S \times 4$ †	$S \times H$
Memory writes	1 * or up to $S \times 5 + 6$ †	0 * or up to H †

* for recall
† for store

At the end of the loop, there is one conditional memory write operation (for zeroing the age of the link). For adding a new link to the map, each iteration of the loop in algorithm 1 includes:

- 1 Euclidean distance measurement operation (for D)
- 2 threshold comparison operations (T and Q)
- 4 memory read operations (3-dimensional vision variable in $M[i]$ and age of $M[i]$)
- 5 conditional memory write operation (remove 3-dimensional value for the vision and 2-dimensional value for the arm in $M[i]$)

At the end of the loop, there are 6 conditional memory write operations (3 for vision, 2 for arm and 1 for the age of the link). Each Euclidean distance measurement operation is composed of:

- 2 sine calculations
- 2 cosine calculations
- 2 subtractions
- 1 addition
- 6 multiplications
- 1 square root

Table 4 is generated for the case where an associative memory stores a set of vector pairs $T = \{X^{(k)}, Y^{(k)}\}_{k=1,\dots,S}$. For the catcher-in-the-rye, $X^{(k)} \in \mathbb{R}^N$ and $Y^{(k)} \in \mathbb{R}^M$, while for the PPAM $X^{(k)}$ and $Y^{(k)}$ are encoded values of the same. The best case values for the catcher-in-the-rye may be obtained by using $S = 1$ while in the worst case S retains its value as the number of elements in the dataset. H in Table 4 represents the number of nodes in the PPAM. Note that each of the H operations execute in parallel in each of the H nodes in the PPAM. Further note that all memory variables in the catcher-in-the-rye are real values and therefore all catcher-in-the-rye calculations require floating point arithmetic.

Despite all these drawbacks of the method used in the catcher-in-the-rye, in terms of sheer speed of operation, it still runs quite fast – primarily because the traditional processors it operates on run at clock frequencies in the range of GHz. However, these traditional processors are produced using components which have become highly refined and suited to the architecture through decades of commercial pressure. On the other hand, digital logic implementations of the PPAM are at a disadvantage and therefore comparing the speed of a PPAM implemented over digital logic with such highly optimised architectures is unfair. Nonetheless, it must be noted that memory recall in the PPAM is asynchronous and, assuming use of efficient electronic components, asynchronous outputs are faster than synchronous outputs.

5.1   Encoding and Quantisation

The catcher-in-the-rye uses an error threshold to control acceptable error levels in the predicted (or recalled) values from the controller. As shown by Hülse et al.

(2010) mappings accumulate uncertainty so that, eventually, the error values distort predictions. For the PPAM, the equivalent of this error threshold is performed implicitly in the encoding. Upon observation of an associative pair, if the PPAM determines that the pair is a new one, it is added to the memory. If, during the encoding process, more values are placed in the same *bin* such that a larger range of values are encoded using the same symbol, the error threshold is increased. However, as shown by Qadir et al. (2011a), unlike the catcher-in-the-rye, minimising this error does not maximise the performance of the PPAM. This (quantisation) error is a parameter that may be tuned.

5.2   Memory Analysis

As shown in Section 4 (Figure 3), the memory capacity of the PPAM increases with the number of conflict resolving nodes. Furthermore, the results in Section 4 indicate that the number of memory locations used in each node is independent of the number of conflict resolving nodes. Together, these two observations imply a high level of scalability since increasing the number of nodes in the network does not effect the structure of individual nodes but it does increase the performance of the PPAM.

   Another concern is the size of the memory required to be implemented in each node. This is dependent upon the width of the data symbols used, since the symbols are used as addresses as well as data in the content addressable memory. The PPAM experiments conducted used 12 bits wide symbols thereby implying an address space of 2^{12} or 4K locations in each node. However, as shown by Figure 5, the maximum number of locations used in any node was 160 while most nodes used far fewer locations. This result agrees with previous observations in Qadir et al. (2010, 2011a) and show that although the theoretical maximum number of calculations in Table 4 may be up to S, the actual number in a real application are far fewer. Therefore, the memory inside nodes in the PPAM can safely be limited to a much smaller number than the theoretical maximum, thereby allowing more nodes to be implemented.

6   Conclusions and Future Work

In this paper, we presented results from experiments performed upon the catcher-in-the-rye dataset (Hülse et al., 2010) that associate the vision space with the reach space of a robot. The results showed that the PPAM is able to store and successfully recall all the data points in the dataset using 29 conflict resolving nodes and can closely approximate this performance using 20 conflict resolving nodes. A comparison of the number and type of operations performed for the PPAM and the catcher-in-the-rye was performed and it was shown that the PPAM utilises the errors from the quantisation process rather than being hindered by it. Finally it was shown that PPAM is scalable and that nodes do not require the theoretical maximum size of the memory to be implemented but rather a much more limited size is enough for memorising data.

Although we have shown in Qadir et al. (2011a) that the PPAM is capable of extracting meaningful relationship information from the data presented to it and therefore can be successfully used to predict values for previously unseen data, these experiments were not performed for this dataset. In future we hope to test this aspect for the catcher-in-the-rye dataset by testing the predicted values and encoding the dataset using different quantisation levels. Furthermore, we hope to repeat the experiments on a real robot–vision system as this would test any implicit assumptions we may have made in the simulation environment.

Acknowledgment. The authors would like to thank the organisers of the Autonomous Intelligent Robots summer school[1] and particularly Martin Hülse who was kind enough to share his data. We would also like to thank our partners at BRL-UWE[2]. The research is funded by the EPSRC funded SABRE[3] project[4] under grant no. FP/F06219211

References

Amaral, J., Ghosh, J.: An associative memory architecture for concurrent production systems. In: Proc. IEEE International Conference on Systems, Man, and Cybernetics 'Humans, Information and Technology', vol. 3, pp. 2219–2224 (1994)

Anzellotti, G., Battiti, R., Lazzizzera, I., Soncini, G., Zorat, A., Sartori, A., Tecchiolli, G., Lee, P.: Totem: a highly parallel chip for triggering applications with inductive learning based on the reactive tabu search. International Journal of Modern Physics C 6(4), 555–560 (1995)

Hebb, D.O.: The Organization of Behavior: A Neuropsychological Theory. Wiley, New York (1949)

Hülse, M., McBride, S., Lee, M.: Robotic hand-eye coordination without global reference: A biologically inspired learning scheme. In: Proc. Int. Conf. on Developmental Learing 2009, China. IEEE Catalog Number: CFP09294 (2009)

Hülse, M., McBride, S., Lee, M.: Fast Learning Mapping Schemes for Robotic HandEye Coordination. Cognitive Computation 2, 1–16 (2010)

Intel: Intel 64 and IA-32 Architectures Optimization Reference Manual. Intel, 248966-024 edition (2011)

Kosko, B.: Bidirectional associative memories. IEEE Trans. Syst. Man Cybern. 18(1), 49–60 (1988)

Lee, P., Costa, E., McBader, S., Clementel, L., Sartori, A.: LogTOTEM: A Logarithmic Neural Processor and its Implementation on an FPGA Fabric. In: International Joint Conference on Neural Networks, IJCNN 2007, pp. 2764–2769 (2007)

Lee, S.W., Kim, J.T., Wang, H.M., Bae, D.J., Lee, K.M., Lee, J.H., Jeon, J.W.: Architecture of RETE Network Hardware Accelerator for Real-Time Context-Aware System. In: Gabrys, B., Howlett, R.J., Jain, L.C. (eds.) KES 2006, Part I. LNCS (LNAI), vol. 4251, pp. 401–408. Springer, Heidelberg (2006)

---

[1] http://users.aber.ac.uk/msh/AIR/

[2] Bristol Robotics Lab University of Western England.

[3] Self-healing cellular Architectures for Biologically-inspired highly Reliable Electronic systems.

[4] http://www.brl.ac.uk/projects/sabre/index.html

Moravec, H.: When will Computer Hardware Match the Human Brain? Journal of Evolution and Technology 1 (1998)

Oh, H., Kothari, S.: Adaptation of the relaxation method for learning in bidirectional associative memory. IEEE Trans. Neural Networks 5(4), 576–583 (1994)

Qadir, O., Liu, J., Timmis, J., Tempesti, G., Tyrrell, A.: Principles of Protein Processing for a Self-Organising Associative Memory. In: Proceedings of the IEEE Congress on Evolutionary Computation (CEC 2010), Barcelona, Spain (2010)

Qadir, O., Liu, J., Timmis, J., Tempesti, G., Tyrrell, A.: From Bidirectional Associative Memory to a noise-tolerant, robust Self-Organising Associative Memory. Artificial Intelligence 175(2), 673–693 (2011a)

Qadir, O., Liu, J., Timmis, J., Tempesti, G., Tyrrell, A.: Hardware architecture for a Bidirectional Hetero-Associative Protein Processing Associative Memory. In: Proceedings of the IEEE Congress on Evolutionary Computation (CEC 2011), New Orleans, USA (2011b)

Teich, J.: Invasive Algorithms and Architectures (Invasive Algorithmen und Architekturen). it - Information Technology 50(5), 300–310 (2008)

Toffoli, T.: CAM: A high-performance Cellular Automaton Machine. Physica D 10, 195–204 (1984)

Turing, A.: Computing machinery and intelligence. Mind 59, 433–460 (1950)

Analysing the Reliability of a Self-reconfigurable Modular Robotic System

Lachlan Murray[1], Wenguo Liu[2], Alan Winfield[2] Jon Timmis[1,3], and Andy Tyrrell[1]

[1] Department of Electronics, University of York, UK
{1jm505,jt517,amt}@ohm.york.ac.uk
[2] Bristol Robotics Laboratory, University of the West of England, UK
[3] Department of Computer Science, University of York, UK

Abstract. In this paper, the reliability of a collective robotic system is analysed using two different techniques from the field of reliability engineering. The techniques, Failure Mode and Effect Analysis (FMEA) and Fault Tree Analysis (FTA), are used to analyse and compare two variants of a previously developed 'autonomous morphogenesis' controller. The reliability of the controller is discussed and areas where improvements could be made are suggested. The usefulness of FMEA and FTA as aids to the design of fault tolerant collective robotic systems, and the comparative effectiveness of the two approaches, is also discussed.

1 Introduction

The SYMBRION project [5] is currently in the progress of developing a large scale heterogeneous collective robotic system. Combining both the swarm and self-reconfigurable modular robotics paradigms, robots may act independently in so-called 'swarm mode' or collectively assemble to form large scale 'organisms'.

A major goal of the SYMBRION project is the long-term survival of collective robotic systems. To which end, the consortium have proposed the '100 Robots 100 Days' grand challenge [4]. As the challenge outlines, for 100 robots to survive without any human interaction for 100 days, they will need to exhibit both high degrees of adaptivity to changes in their environment and extreme tolerance to the presence of faulty individuals. The work here focusses on the latter.

By applying Failure Mode and Effect Analysis (FMEA) to the controller of a swarm robotic system, Winfield and Nembrini [11] showed that although in many cases such systems do display high degrees of fault tolerance, in certain scenarios, the effects of a single failure may be catastrophic. Building upon this work, we apply the same technique to the collective robotic system of SYMBRION. We also apply a second contrasting technique from the field of reliability engineering, Fault Tree Analysis (FTA). Two variations of the morphogenesis controller originally developed in [7] are analysed. The controller is introduced in section 2. In sections 3 and 4 FMEA and FTA are performed. In section 5 the outcomes of the analysis are discussed and the FMEA and FTA procedures are compared. Conclusions and potential avenues of future work are presented in section 6.

E. Hart et al. (Eds.): BIONETICS 2011, LNICST 103, pp. 44–58, 2012.

2   Morphogenesis Controller

The morphogenesis controller of [7] was designed to allow a swarm of robots to self-assemble into an artificial organism of predetermined size and shape. Once assembled, at a later stage, the organism may either completely disassemble, returning every module to swarm mode, or partially disassemble, allowing for the efficient transformation into a different shape. Before describing the controller itself, it is necessary to briefly introduce the robots for which it was developed.

SYMBRION, and the associated REPLICATOR project, are currently developing three unique, yet complimentary, robotic platforms [4]. Here, we are concerned primarily with the so-called 'Backbone' platform. To aid in the formation of organisms, Backbone robots possess precise omnidirectional 2D locomotion, and four docking sides at which other robots may connect. Each docking side contains a set of infrared (IR) components which allows robots to locate and align with each other, a docking element that allows them to physically connect, and an interface through which they may transfer both energy and data.

The docking elements contain a mechanism which physically locks the connection between two robots in place. For a reliable connection to be established between two robots, at least one of the pair must have their element locked.

The IR subsystems that help robots to align with each other are critical to the performance of the morphogenesis controller. Together, the various IR components serve a variety of different functions. Specifically, on each docking side of a robot there are: two IR sensors, for obstacle and robot proximity detection; three IR LEDs, which allow the robot to act as a beacon or broadcast messages to its neighbours; and a single IR remote receiver, for detecting the messages sent by others. For full details of how these components are integrated see [6].

Every robot in the system runs the same behavioural controller, a finite state machine (FSM) for which is shown in figure 1. The behaviours labelled with an 'O' in figure 1 are executed by modules that are currently in the organism, and the behaviours labelled with an 'S' are executed by robots in swarm mode.

Figure 1 shows the behaviours of the individual robots. It is also possible to identify 'system level' behaviours which arise from the *interactions* of individual robots. At least three such system level behaviours may be identified: *Exploration*, *Self-assembly* and *Self-disassembly*. These three behaviours together allow the system to exhibit the property of autonomous morphogenesis.

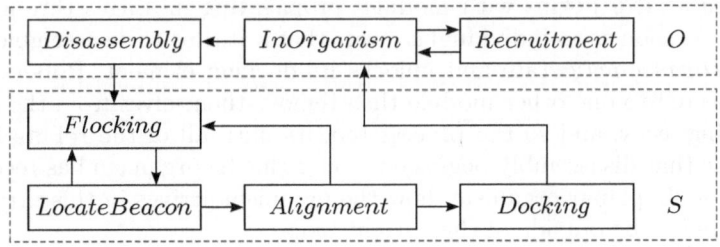

Fig. 1. A finite state machine for the morphogenesis controller of [7]

Exploration is provided by the interactions of the robots in the *Flocking* state. *Flocking* is a place holder for any swarm mode behaviour [7], in this scenario it may more accurately be described as 'obstacle avoidance'. Robots simply move forward and if an obstacle is encountered, avoid it. In an enclosed arena with no internal obstacles, this strategy is sufficient to ensure good coverage.

Self-assembly begins with robots in the *Recruitment* state broadcasting both long-range 'recruitment' messages and short range beacon signals. The detection of 'recruitment' messages causes robots in the *Flocking* state to transition into the *LocateBeacon* state and subsequently, if the beacon signal is detected, into the *Alignment* state. After alignment, in order to establish a reliable connection, a simple docking protocol is executed. At the same time as it locks its own docking element, the docking robot instructs the recruiting robot to do the same by sending a 'docking-ready' message using its IR LEDs. After a short delay, allowing time for the connection to be established, the docking robot transmits a 'docking-complete' message through the wired channel. The recruiting robot responds by transferring information regarding the shape of the current organism.

When a new robot joins the organism it enters the *InOrganism* state. When a module is require to recruit it transitions to *Recruitment*, and when finished transitions back to the *InOrganism* state. The decision as to which robots enter the *Recruitment* state at which point in time, and upon which sides they signal for other robots to join, is determined by a 'recruitment strategy'. Two different recruitment strategies have been developed. The first of which is described in [7] and referred to as '*single entry recruitment*'. In this strategy only one robot may occupy the recruitment state at any moment in time, and may only recruit upon one docking side. The strategy relies on modules propagating messages whenever a new robot joins, so that every robot is aware of the state of completion, and hence when it is their turn to recruit. The second strategy, currently in press [8], is referred to as '*multiple entries recruitment*'. This strategy removes the restrictions upon the number of robots that may recruit simultaneously, and the number of sides at which they may do so. As soon as a robot joins the organism, if it is required to recruit, it immediately transitions to the *Recruitment* state.

The controller also incorporates precautions against the interference that may occur as multiple robots attempt to align with the same recruiting module. For example, whilst aligning, robots broadcast 'expelling' messages, which, if detected by another robot, cause that robot to switch back to *Flocking*.

Self-disassembly begins with modules propagating messages throughout the organism. As soon as an individual is aware that it is time to disassemble it moves into the *Disassembly* state and unlocks its docking element. Robots that are only connected to one other module then remove themselves from the organism by reversing away, and so the process repeats until all of the robots have left. We assume that disassembly begins only once the 3D organism has reconfigured itself into a 2D planar organism, how the organism arrives in this arrangement is considered to be outside of the current scope.

3   Failure Mode and Effect Analysis

Used widely throughout the manufacturing industry, Failure Mode and Effect Analysis (FMEA) is a well established procedure for analysing the safety and reliability of a product or process [9]. For every component in the system under study, the analyst will derive a list of specific 'failure modes' and then attempt to identify all of the effects that these failures may have on the system, in the process building up a general overview of the system's reliability. Because of this progression from specific failures to general effects, FMEA may be described as an *inductive* approach to system analysis. For further details refer to [2] and [9].

To perform FMEA on the morphogenesis controller we follow a similar approach to Winfield and Nembrini [11]. The individual components of the system are considered to be the system level behaviours introduced in section 2. The failure modes or 'hazards', meanwhile, correspond to the complete or partial failure of an individual robot. Like [11], we consider only internal hazards. The six hazards identified by [11] (reproduced in table 1a) form the basis of this work.

Table 1. Hazards investigated by [11] (a) and those analysed in this study (b)

Hazard	Description		Hazard	Description
H_1	Motor failure		H_M	Motor failure
H_2	Communications failure		H_R	IR remote receiver failure
H_3	Avoidance sensor(s) failure		H_S	IR sensor failure
H_4	Beacon sensor failure		H_L	IR LED failure
H_5	Control systems failure		H_T	Total systems failure
H_6	Total systems failure		H_D	Docking element failure
			H_W	Wired communication failure

(a)	(b)

The majority of the hazards considered in [11] correspond to the failure of an independent subsystem. In this study, the subsystems responsible for communications, avoidance sensing and beacon sensing are all inter-linked. As shown in table 1b, because of this lack of independence, hazards corresponding to the low level components that make up each subsystem, rather than the subsystem itself, are considered. Motor and total systems failures are kept without alteration.

In section 2 three system level behaviours were identified: exploration, self-assembly and self-disassembly. When an individual robot suffers a failure, the effect that it has on the system will differ depending upon which of the system level behaviours the robot was contributing to when the failure occurred. During self-assembly, the choice of recruitment strategy will influence the effects of any failures. Because we are interested in comparing the two recruitment strategies, during analysis we extract the recruitment strategy decision process from the self-assembly behaviour and consider it as a separate system level behaviour.

Whilst performing FMEA, four different effects were identified, two of which are described as serious and two of which are described as non-serious. All four are listed below, with uppercase lettering used to denote the serious effects:

e_1 - *reduction in the number of capable robots*

e_2 - *delay in the formation of an organism*

E_1 - *stall in the formation of an organism*

E_2 - *stall in the disassembly of an organism*

The various scenarios in which these effects occur are now outlined. The full analysis of all seven hazards is too long to reproduce in full. Consequently, although summaries are provided, details of the effects of hazards H_L and H_R, which were identified as being the least detrimental to the system, are omitted.

H_M - Motor Failure. A robot that has suffered a motor failure will remain stationary. It will not, however, be prevented from communicating with others.

The effect that a motor failure has on the system depends largely upon where the robot was located when the failure occurred. The failure of a robot that is located outside of communication range of the organism, thus contributing only to exploration, will have very little effect on the system. Since there is no explicit communication between members of the swarm during exploration, a failed robot will not interfere significantly with others. The only effect will be a reduction in the number of capable robots, e_1.

Robots in swarm mode which are contributing to the self-assembly behaviour will be located closer to the organism. The effect of a motor failure in this instance will be far more severe. A robot that stops moving whilst in the *Alignment* state, for example, will not only fail to join the organism itself, but by transmitting 'expelling' messages and physically blocking the path of others, may actively prevent further robots from doing so. If robots are prevented from joining the organism at one or more recruitment sites, the formation of the organism will be permanently stalled, effect E_1.

Analogously, during self-disassembly, a robot that has suffered a motor failure may act as a physical obstacle, preventing itself and others from leaving the organism, resulting in a stall in the disassembly of an organism, or effect E_2.

H_S - Infrared Sensor Failure. IR sensors can fail in a variety of different ways. In this study, no single type of failure is considered, but rather the focus is on the general effects of a breakdown in the correlation between true and sensed values. IR sensors are used for both proximity detection and beacon detection. The failure of one or more IR sensors will reduce the robot's ability to perform both of these functions. Since the recruitment strategies do not require these functions they will not be affected, the other behaviours however, will be.

Without the ability to effectively perform proximity detection, a robot is likely to collide with obstacles or its neighbours. This is not a problem if the robot is contributing only to exploration, in which the effect will simply be a reduction in the number of capable robots, effect e_1. If a robot is contributing to self-assembly or self-disassembly, however, the inability to perform proximity detection may lead to the robot colliding with the organism. By physically blocking the path of other robots, in the worst case, an IR sensor failure may cause both the assembly and disassembly processes to stall, effects E_1 and E_2.

Without the ability to perform beacon detection, robots will be unable to align with the recruiting module, this may simply lead to the robot returning to the *Flocking* state and the exploration behaviour, but in the worst case it may lead to the robot colliding with the organism, again effect E_1.

H_T - Total Systems Failure. A total systems failure —which may be considered equivalent to a robot running out of energy— will result in the shutdown of all of a robot's subsystems. A robot that suffers a total systems failure will be immobilised and unable to communicate with other robots.

During exploration the effects of a total systems failure are negligible. Any robot contributing to exploration that suffers a total systems failure will simply act as a static obstacle to other members of the swarm. The only effect will be a reduction in the number of active robots, e_1.

For similar reasons as to when a motor failure occurs, a robot that suffers a total systems failure whilst blocking the path to or from an organism, may cause both the assembly and disassembly of the organism to stall, effects E_1 and E_2. Note, however, that unlike the effects of H_M there will be no interference due to the robot sending 'expelling' messages. A total systems failure will also prevent recruiting modules from attracting new robots and disassembling modules from unlocking their docking mechanisms, further precursors of effects E_1 and E_2.

Robots which fail whilst executing the recruitment strategy will be unable to transition from the *InOrganism* state to the *Recruitment* state when required. Furthermore, in the case of single entry recruitment, will be unable to propagate messages when a new robot joins the organism. Again leading to effect E_1.

H_D - Docking Element Failure. The failure of a docking element will force the mechanism to remain in the state that it currently occupies. Docking elements do not change state during exploration, or during the execution of either recruitment strategy, their failure in these behaviours, therefore, will have no effect.

Assuming that all robots start with their docking elements unlocked and remembering that the docking protocol introduced in section 2 ensures that only one locked docking element is required for a reliable connection to be established, if a single element fails in the unlocked state, self-assembly will not be affected.

If the docking element of an existing member of the organism fails-locked, however, it will prevent the affected robot (and any connected neighbours) from un-docking. Thus leading to effect E_2, a stall in the disassembly of the organism.

H_W - Wired Communication Failure. The failure of wired communications will prevent the affected robot from sending or receiving messages. Furthermore, the failed robot will be unable to act as a node in a wired network. Not only will the failure have a local effect on the failed robot, but any two robots in the organism whose sole path of communication passes through the failed module will also be prevented from exchanging messages. Wired communications are not utilised during exploration, nor are they used during the decision process of the multiple entries strategy, in these scenarios such a failure will have no effect.

During self-assembly, the failure of wired communications in either a recruiting robot or a newly docked robot, will prevent the new module from receiving information about the shape of the organism being constructed. Thus, the assembly of the organism will stall, effect E_1.

During self-disassembly, if a module is unable to propagate messages when it is time to disassemble, sections of the organism may be unaware that it is time to do so. Resulting in a stall in the disassembly of the organism, or effect E_2.

Similarly, in order to determine when a robot should enter the *Recruitment* state, the single entry recruitment strategy requires that messages are propagated throughout the organism when a new module joins. If this is not possible then the formation of the organism will stall with effect E_1.

The effects that all seven of the hazards may have on the performance of the system are summarised in table 2. The three system level behaviours and the two recruitment strategies are considered separately.

Table 2. FMEA Summary

Behaviour	H_M	H_R	H_S	H_L	H_T	H_D	H_W
Exploration	e_1	e_1	e_1	-	e_1	-	-
Self-assembly	E_1	e_2	E_1	E_1	E_1	-	E_1
Self-disassembly	E_2	-	E_2	-	E_2	E_2	E_2
Recruitment strategy							
Single entry	-	-	-	-	E_1	-	E_1
Multiple entries	-	-	-	-	E_1	-	-

As shown in table 2, of the 21 possible combinations of hazard and behaviour (ignoring for now the recruitment strategies), all but 6 will have some effect on the system. Of the 15 which have an effect, 10 of the effects are described as serious. The most detrimental of the seven hazards is a total systems failure, which has a negative effect on the system in every behaviour, as well as during the execution of both recruitment strategies. The failure of an infrared remote receiver is the least harmful, in no scenario will this hazard have a serious effect on the system. LED failures and docking element failures are the next least critical, each only leading to a single serious effect, in a single behaviour.

In comparing the two recruitment strategies, the only difference that arises is the effect of a wired communications failure. Systems employing both strategies are reliant upon wired communications to ensure that newly docked robots receive information about the shape of the organism being constructed. After a robot has joined the organism, however, only the single entry strategy uses wired communications for determining which robots should enter the *Recruitment* state. With the single entry strategy the failure any robot that prevents messages being propagated to the required modules will cause the formation of the organism to stall. With the multiple entries strategy, meanwhile, only a failure during the initial exchange of information will stall the formation of an organism.

4 Fault Tree Analysis

In contrast to FMEA, Fault Tree Analysis (FTA) [10] is a *deductive* approach to failure analysis. Beginning with a general system failure, the analyst attempts to identify all of the potential causes of the event and the logical sequence of secondary events leading up to it, these together make up the fault tree.

The tree for a particular system failure is constructed by repeatedly breaking the event down into more specific intermediate events. The relationships between events at different levels in the tree are specified using Boolean logic, where the outputs of functions at lower levels serve as the inputs to those at higher levels.

Following the construction of a fault tree, and relying on the fact that the underlying structure of the tree may be described in terms of Boolean algebra, the analyst may perform both qualitative and quantitative analysis (this again differentiates FTA from FMEA, in which only qualitative analysis is possible). Qualitatively, the analyst is able to identify the various combinations of component failures which may cause a system failure, as well as their relative importance. Quantitatively, if the failure rates of the individual components are known, it is possible to calculate the probability of a system failure.

In the remainder of this section, a simple fault tree is constructed for the morphogenesis controller. The tree is used to analyse some of the potential causes of a stall in the formation of an organism. As we are limited by space, however, only a small part of the entire fault tree for this event is presented. Fault trees are represented graphically using an extension of standard logic gate notation. In this work, only a subset of the available notation is used, with symbols introduced as and when they are required, for a full list see [10]. Following construction, the tree is analysed qualitatively. The combinations of component failures which may lead to a stall in the assembly process are identified and their relative importance discussed. Comments are made upon the levels of fault tolerance in the current system and areas in which this could be improved are identified.

Construction. Before beginning construction we must state more precisely what is meant by a "stall in the formation of an organism". A stall in the formation of the organism can be said to have occurred if at least one robot has been unable to recruit the modules that it was required to (within some acceptable time limit). This means that one or more robots have either become stuck in, or failed to transfer to, the *Recruitment* state. When investigating the multiple entries recruitment strategy we need only to consider the scenario in which a robot becomes stuck in the *Recruitment* state, since the time spent in the *InOrganism* state (before a robot has completed its recruiting duties) is negligible. With the single entry strategy we must consider both cases. Here we examine only the first scenario, that a robot has become stuck in the *Recruitment* state.

The only condition that will cause a robot to leave the *Recruitment* state is the receipt of confirmation (through wired communications) that another robot

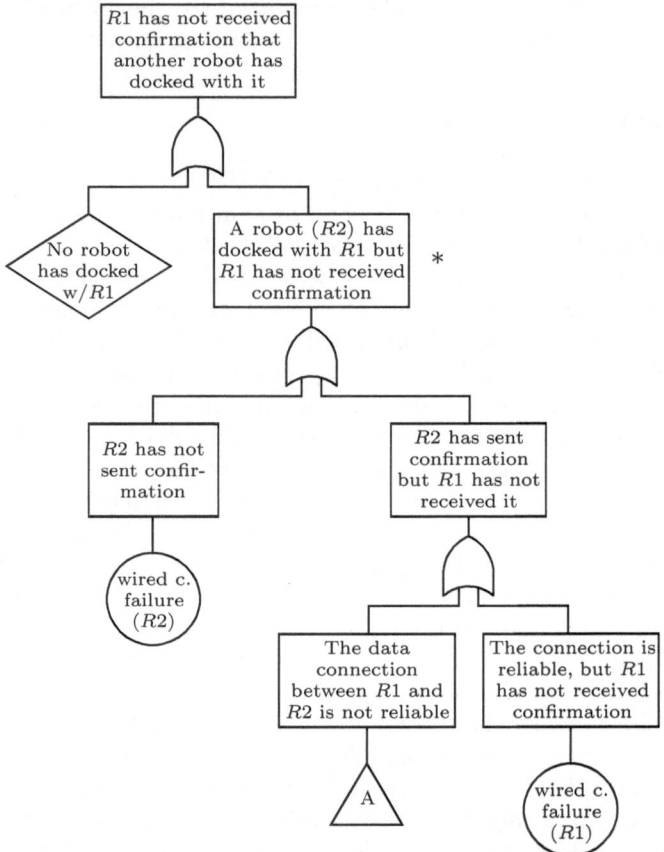

Fig. 2. Fault tree examining the causes of a stall in the formation of an organism

has docked with it. As shown in figure 2, where $R1$ is the robot that is awaiting confirmation, this statement constitutes the topmost event of our fault tree. In fault tree notation, events are represented using rectangular boxes.

We now consider the reasons why $R1$ has not received confirmation. There are two options, either no robot has docked with $R1$, or a robot ($R2$) has docked with $R1$ but for some other reason $R1$ has not received confirmation. These options are separated using an OR gate, representing the fact that the output of this gate will occur if at least one of the inputs occurs. The scenario in which no robot has docked with $R1$ is not pursued further, and as such this 'undeveloped event' is presented in a diamond shaped box. We now focus on the case in which another robot has docked with $R1$. In this context, we consider two robots to have 'docked' once they have aligned and moved into the docking position, communication between the pair and the engagement of either docking mechanism are not considered necessary for docking to be said to have taken place. There are two immediate reasons why $R1$ would not receive confirmation of docking from $R2$. The first is that $R2$ has not sent confirmation, the second

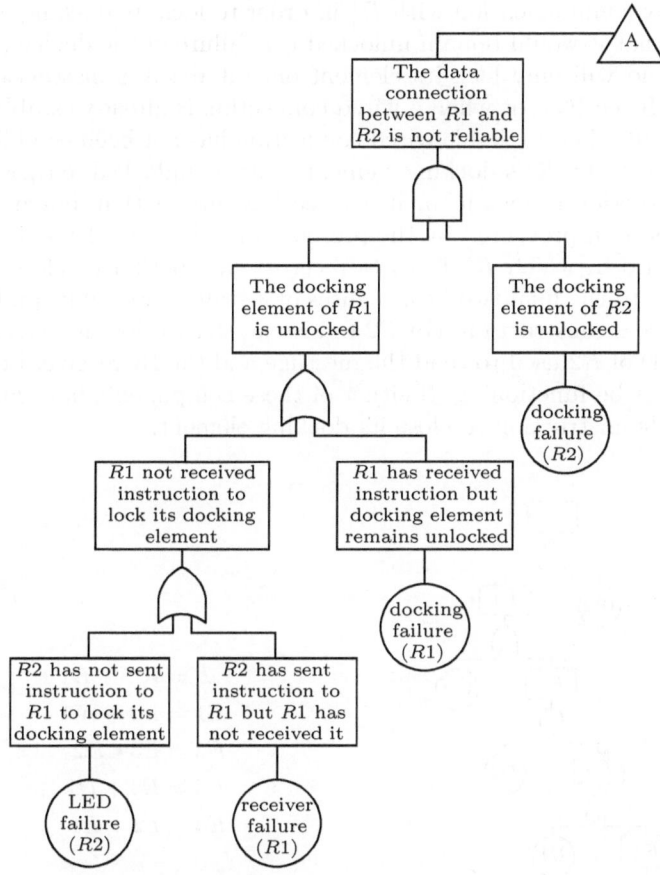

Fig. 3. Fault tree examining the causes of a stall in the formation of an organism

is that $R2$ has sent confirmation but $R1$ has not received it. Considering the first leads to the first basic cause of a stall in the assembly of an organism, that $R2$ has suffered a wired communications failure. Such 'basic fault events' are represented by circles. The second case, that $R2$ has sent confirmation but $R1$ has not received it, requires further development.

There are two possibilities why, although $R2$ has sent confirmation, $R1$ has not received it. The first is that the data connection between the two robots is unreliable. The second is that although the data connection is reliable, $R1$ has still not received confirmation. In this second scenario, given that confirmation was successfully sent by $R2$, the problem must lie with the wired communications of robot $R1$. The case of an unreliable data connection is developed further in figure 3, triangular 'transfer symbols' are used to connect the two diagrams.

Using an AND gate, the output of which will occur only if both inputs do, the first part of figure 3 shows that for an unreliable connection to be formed between $R1$ and $R2$, the docking elements of both robots must be unlocked. Remembering the docking protocol introduced in section 2. Since $R2$ is not reliant on any communication with $R1$ in order to lock its docking element, the only reason that it would remain unlocked is a failure of the device itself. $R1$ on the other hand will only lock its element once it receives instruction from $R2$ through an IR receiver, or when a wired connection is already established. Since we are assuming that a reliable wired connection has not been established, there are two reasons why $R1$'s docking element remains unlocked, either $R1$ has not received instruction to lock it, or it has received instruction, but in attempting to lock it has been prevented by the presence of a docking element failure.

The scenario in which $R1$ has not received instruction to close its docking element leads to the final two basic causes of a stall in assembly, and to the end of the fault trees construction. For $R2$ to instruct $R1$ to close its docking element both the LED of $R2$ used to send the message and the IR receiver of $R1$ used to receive it must be functioning. If either of these components have failed $R1$ will not receive the instruction to close its docking element.

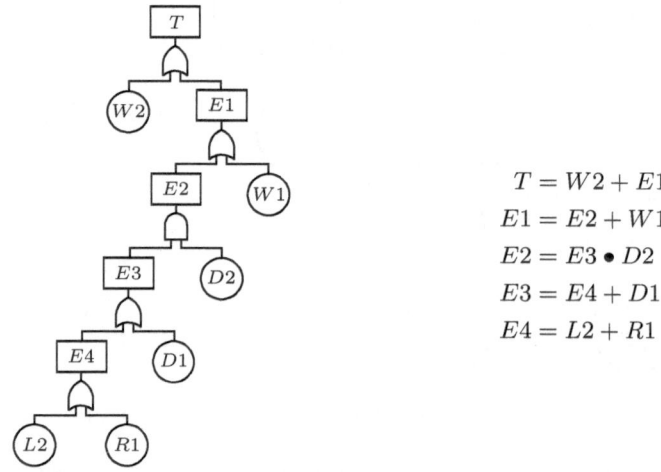

$$T = W2 + E1$$
$$E1 = E2 + W1$$
$$E2 = E3 \bullet D2$$
$$E3 = E4 + D1$$
$$E4 = L2 + R1$$

Fig. 4. A minimal version of figures 2 and 3, with equivalent boolean equations

Analysis. The first step of analysing a fault tree involves representing it in terms of boolean equations. Ignoring the undeveloped event, figures 2 and 3 can be minimally represented as figure 4. Where the basic fault events WX, DX, LX and RX correspond respectively to wired communications, docking element, LED, and receiver failures of robot 'X'. The equivalent boolean equations of the minimal tree are shown along side it in figure 4.

With the tree converted into Boolean equations, the next task is to obtain the "minimal cut sets". A minimal cut set is any combination of basic events which, if they all occur, will cause the top event to occur. The minimal cut sets can be obtained simply by reducing the equations of the fault tree until only basic fault events are left. The equations in figure 4 are easily converted into:

$$T = (D1 \bullet D2) + (D2 \bullet L2) + (D2 \bullet R1) + W1 + W2$$

The minimal cut sets then, are simply: $\{D1, D2\}$, $\{D2, L2\}$, $\{D2, R1\}$, $\{W1\}$ and $\{W2\}$. Where, for example, $\{D1, D2\}$ represents the scenario in which both a recruiting module and a newly docked robot both suffer docking element failures.

For this simple example, deriving the minimal cut sets was trivial. However, it should be remembered that this is only a subsection of a much larger fault tree. To determine the minimal cut sets of more complex fault trees requires greater effort. Fortunately, software exists to aid the analyst during both the construction and analysis of fault trees. One example of FTA software which was used to help develop the fault trees in this work is OpenFTA [1].

Even though we have only considered a subsection of the overall tree, it is still possible to make some interesting qualitative observations. From the point of view of reliability, single component minimal cut sets are undesirable, simply because they imply the failure of a single component may cause the entire system to fail. Even without reducing figure 4 to the minimal cut sets of its Boolean equations it is obvious from the lack of AND gates that the system will contain single component cut sets. In this case, those cut sets are $\{W1\}$ and $\{W2\}$, corresponding to the scenarios in which either a recruiting robot or a newly docked robot has suffered a wired communications failure. Figure 4 does contain one AND gate, and its presence gives rise to the three remaining two-component cut sets, all of which it may be noted, contain $D2$. The design of the hardware and the simple docking protocol, in this situation, ensure that the single failure of one of the involved components does not lead to a stall in the assembly process.

As a general rule, it is desirable to have AND gates positioned as high up the fault tree as possible. With this in mind it is possible to suggest some improvements to the system. A weak point in the fault tree of figure 2 may be identified at the '∗'. Although recruiting modules are provided with information about the presence of newly docked robots in the form of both wired and wirelessly transmitted messages, as well as the values returned by their IR proximity sensors, they are critically dependent upon the wired channel to determine their behaviour. Recall that the only time at which a robot will leave the *Recruitment* state is when it receives confirmation, through wired communications, that another robot has docked with it. It is suggested, therefore, that rather than relying solely on wired communications to dictate their behaviour, recruiting modules should make better use of the data from their IR receivers and sensors. If robots react appropriately to the information (or lack thereof) that all three channels provide, the chances of a robot becoming stuck in the recruitment state, following the docking of another robot, may be severely reduced. For example, detecting the presence of another robot without receiving confirmation through wired communications may signal to the recruiting robot that there is something wrong. Thus providing the robot with another way out of the recruiting state by, for example, initiating the processes of repairing or replacing either itself or the docking robot. The fault tree of a controller adapted in this manner, no longer being singularly reliant on wired communications, would be expected to contain more AND gates, located further up the tree. The system, therefore, would exhibit greater tolerance to the failure of individual components.

5   Discussion

Whereas FMEA is an inductive approach to failure analysis, FTA is deductive. When performing FMEA, the analyst begins with a list of specific hazards and attempts to identify what general effects they may have on the system. Conversely, with FTA, the analyst begins with a general effect and gradually traces it back to its root causes, the specific hazards. In practical terms, this means that by employing FTA, the analyst is forced to focus on identifying increasingly specific causes. Resulting in a very complete understanding of the chain of events that lead from a component failure to a system failure.

One advantage of the precise knowledge that the FTA procedure provides, is that it may help the analyst to identify specific weak points in the system. This was shown in section 4 where the over reliance of recruiting robots on wired communications was identified as a weakness. What the FMEA procedure lacks in terms of its depth of detail, it makes up for with the breadth of information that it provides. In section 3 the FMEA procedure covered three different effects of five different single component failures. In a similar amount of space, with significantly greater effort, the FTA procedure covered only a subsection of a single effect, and revealed only one single component cause.

What the FTA procedure does reveal with ease, which may be less obvious when performing FMEA, are the different *combinations* of component failures which may lead to a system failure. Furthermore, although in general, FMEA is better suited to the task of exhaustively listing possible hazards, the two approaches both have the potential to identify hazards overlooked by the other.

Another advantage of FTA, which was not exploited in this work, is the ability to perform quantitative analysis. One of the reasons why quantitative analysis was not carried out is because it is reliant upon failure rate data, which was not available. Another more pressing reason, however, indicative of a much larger problem with the application of FTA to collective robotic systems, is that the systems themselves may be too complex to model in sufficient detail. When constructing the fault trees here, assumptions were made about the state of the system, in particular with regard to the interactions between robots and their environment. The tree presented in section 4 examined the system in a relatively static state, considering the interactions between only two robots. Other, more dynamic parts of the system require increasingly limiting assumptions. Allowing for a more complete understanding, but at the expense of detail in the model. Without modelling the interactions between robots accurately, it is not possible to produce accurate quantitative results. The dynamic nature of these systems, as well as the large numbers of agents and the locational dependency of their interactions make such systems difficult to model using standard fault trees.

The desire to model the reliability of complex fault tolerant systems is widespread. The difficulties that arise in accounting for the high levels of redundancy, fault recovery mechanisms and sequentially dependent failures that such systems possess are not unique to robotic systems. Efforts to help solve these problems have led to the development of Dynamic Fault Trees (DFT) [3]. In constructing DFTs, additional 'dynamic' gates are used to account for the extra complexity. Analysing DFTs then essentially involves combining the standard fault tree approach with Markov Chain models. Thus, the simplicity of the standard approach is augmented by the flexibility of Markov models [3].

The analysis carried out in section 3 revealed that 15 of the 21 possible combinations of hazard and behaviour will have some effect on the system, 10 of which will be serious. This picture is stark in contrast to the results of the study by Winfield and Nembrini [11], in which there were observed to be only 6 serious effects out of a possible 30. While performing FMEA, a pessimistic view was taken in assuming that all components of a single subsystem will fail simultaneously. Whilst in some cases, the effect of a failure will depend upon how many and which components of a subsystem fail, in others, the failure of a single component may be equally as detrimental as the failure of multiple. The poor outlook, therefore, cannot be attributed solely to pessimism. Nor should we be quick to denounce the long held belief that swarm robotic systems inherently provide fault tolerance, even if in certain scenarios, as shown by [11], its foundations are questionable. We are not dealing with a pure swarm robotic system, but with a combined swarm and modular robotic system. The modular aspect brings with it far greater dependence between robots. If we examine the system closer, we see that it is during the interactions with modules already in the organism where the effects of a failure are most severe. Note that in table 2 there are no serious effects when a failure occurs during exploration, the only behaviour that involves robots operating exclusively in swarm mode.

To endow a modular robotic system with levels of fault tolerance similar to those found in swarm robotic systems, greater plasticity in the conformation of the system is required. Whereas in swarm robotic systems, faulty robots may simply be 'left behind', in modular robotic systems, more explicit methods of replacing or repairing modules are required. The SYMBRION robotic platforms were designed with this in mind, supporting the ability of robots to recharge or directly power their connected neighbours. As a result, in our pursuit to design fault tolerant modular robotic systems, we may be encouraged by the fact that what appears to be the most detrimental hazard, a total systems failure, is also the easiest to repair (if assumed to be caused by a robot running out of energy).

In this analysis we did not consider the effects of either transient or external hazards. If transient hazards are considered, or if it is possible to repair or replace failed modules, the multiple entries strategy possesses a significant advantage. With the multiple entries strategy, even though the assembly of an organism may stall at one point, it may continue at others. A system utilising the multiple entries strategy is therefore able to recruit and repair simultaneously. In considering external hazards, such as the interference observed near recruitment sites, the single entries recruitment strategy possesses the advantage. However, with the multiple entries strategy, the increased level of interference is offset by the fact that multiple robots may recruit simultaneously. The time saved by the multiple entry strategy directly translates to a saving in energy, which may be critical to the long-term survival of the system, and consequently, of great importance to the completion of the '100 Robots 100 Days' grand challenge.

6   Conclusions and Future Work

FMEA and FTA have been applied to the morphogenesis controller of a combined swarm and self-reconfigurable modular robotic system. The analysis revealed several scenarios in which even a single failure may have serious consequences, the worst case being the occurrence of a total systems failure. FMEA and FTA were each identified to have their own advantages. Whereas FMEA provides a good general overview, FTA reveals specific details about the chain of events that may lead to a system failure. A combined approach, as demonstrated here, is advocated during the design of fault tolerant collective robotic systems.

Using the knowledge gained in this study, future work will investigate ways of improving the fault tolerance of the morphogenesis controller. The focus will primarily be on addressing total systems failures, particularly within the scope of energy management. Further analysis of the both the current controller and any improved controllers, may also be performed. For which purpose, DFTs and Markov chain models have been identified as a promising future prospects.

Acknowledgments. The SYMBRION project is funded by the European Commission, within the 7th Framework Programme. Project No. FP7-ICT-2007.8.2.

References

1. Auvation. OpenFTA [Computer software] (2011), http://www.openfta.com
2. Dailey, K.W.: The FMEA Pocket Handbook. DW Publishing Co. (2004)
3. Dugan, J., Bavuso, S., Boyd, M.: Dynamic fault-tree models for fault-tolerant computer systems. IEEE Transactions on Reliability 41(3), 363–377 (1992)
4. Kernbach, S., Scholz, O., Harada, K., Popesku, S., Liedke, J., Raja, H., Liu, W., Caparrelli, F., Jemai, J., Havlik, J., Meister, E., Levi, P.: Multi-Robot Organisms: State of the Art. In: ICRA 2010, Workshop on "Modular Robots: State of the Art", Anchorage, pp. 1–10 (2010)
5. Levi, P., Kernbach, S.: Symbiotic Multi-Robot Organisms: Reliability, Adaptability, Evolution. Cognitive Systems Monographs. Springer (2010)
6. Liu, W., Winfield, A.F.T.: Implementation of an IR approach for autonomous docking in a self-configurable robotics system. In: Kyriacou, T., Nehmzow, U., Melhuish, C., Witkowski, M. (eds.) Proceedings of Towards Autonomous Robotic Systems, pp. 251–258 (September 2009)
7. Liu, W., Winfield, A.F.T.: Autonomous Morphogenesis in Self-assembling Robots Using IR-Based Sensing and Local Communications. In: Dorigo, M., Birattari, M., Di Caro, G.A., Doursat, R., Engelbrecht, A.P., Floreano, D., Gambardella, L.M., Groß, R., Şahin, E., Sayama, H., Stützle, T. (eds.) ANTS 2010. LNCS, vol. 6234, pp. 107–118. Springer, Heidelberg (2010)
8. Liu, W., Winfield, A.F.T.: Distributed autonomous morphogenesis in a self-assembling robotic system. In: Doursat, R., Sayama, H., Michel, O. (eds.) Morphogenetic Engineering: Toward Programmable Complex Systems (to appear)
9. McDermott, R.E., Mikulak, R.J., Beauregard, M.R.: The Basics of FMEA. Productivity Press (2008)
10. Vesely, W.E., Roberts, N.H.: Fault Tree Handbook. U.S. NRC (1981)
11. Winfield, A., Nembrini, J.: Safety in numbers: fault-tolerance in robot swarms. International Journal of Modelling, Identification and Control 1(1), 30–37 (2006)

BIO-CORE: Bio-inspired Self-organising Mechanisms Core

Jose Luis Fernandez-Marquez[1], Giovanna Di Marzo Serugendo[1],
and Sara Montagna[2]

[1] University of Geneva, Switzerland
joseluis.fernandez@unige.ch, giovanna.dimarzo@unige.ch
[2] Università di Bologna
sara.montagna@unibo.it

Abstract. This paper discusses the notion of "core bio-inspired services" - low-level services providing basic bio-inspired mechanisms, such as evaporation, aggregation or spreading - shared by higher-level services or applications. Design patterns descriptions of self-organising mechanisms, such as gossip, morphogenesis, or foraging, show that these higher-level mechanisms are composed of basic bio-inspired mechanisms (e.g. digital pheromone is composed of spreading, aggregation and evaporation). In order to ease design and implementation of self-organising applications (or high-level services), by supporting reuse of code and algorithms, this paper proposes BIO-CORE, an execution model that provides these low-level services at the heart of any middleware or infrastructure supporting such applications, and provides them as "core" built-in services around which all other services are built.

Keywords: Bio-inspired design patterns, self-organising systems' engineering.

1 Introduction

The current situation in design and development of bio-inspired or decentralised systems can be compared to the situation some 40 years ago when programs were written in the assembly language. To compute an addition, using the assembly language, we need to move both operands into appropriate registers and then apply the addition operator. Today, to implement a system exploiting ant foraging using pheromones, in addition to programming the foraging behaviour, it is also necessary to implement the behaviour of the pheromone itself. An additional inconvenience today with bio-inspired systems resides in the fact that if two applications use pheromones, they both need to implement their own version of the pheromone even though the two applications run in the same node. What we need is the possibility to program bio-inspired systems using high-level operators manipulating core mechanisms as "first-class entities" in a way similar to how high-level programming languages helped abstracting away the implementation of the addition for the programmer and thus favoured the re-use

E. Hart et al. (Eds.): BIONETICS 2011, LNICST 103, pp. 59–72, 2012.
© Institute for Computer Sciences, Social Informatics and Telecommunications Engineering 2012

of code. Additionally, as the same addition operator can be used for many different applications, the same pheromone mechanism can be shared by different applications.

Bio-inpired design patterns provide solutions for existing recurrent problems. A design pattern clearly identifies the problem that a mechanism solves, where it has been applied and what are the consequences or emergent behaviour we can observe after applying the pattern. Identifying the bio-inspired mechanisms and their boundaries is a first step towards systematic design and development of self-organising systems. However, we are still far from high-level programming, where designers and programmers concentrate on which mechanisms they want to use, how they want them to be combined, relying on a middleware for the actual distributed implementation of these mechanisms.

In our previous work [Fernandez-Marquez et al., 2011b, Fernandez-Marquez et al., 2011a], we focused on relations between mechanisms and presented several self-organising mechanisms under the form of design patterns, clearly identifying the boundaries of each mechanism, showing how the different patterns relate with each other and how some patterns are composed by others. We classified a set of bio-inspired mechanisms into three different layers: (1) Basic Mechanisms are those mechanisms that cannot be further decomposed and are used to compose other mechanisms or used alone; (2) Composed mechanisms are those mechanisms composed by basic mechanisms; and (3) Top-level mechanisms are those mechanisms exploiting basic and composed ones. From this classification we concluded that most of the bio-inspired patterns are actually using basic bio-inspired mechanisms. Additionally, these basic mechanisms are usually executed by the environment when we consider biological processes (e.g. spreading, aggregation and evaporation of pheromones).

This paper presents BIO-CORE, an execution model for "core bio-inspired services", providing basic bio-inspired mechanisms as built-in services. Such a core set of services, typically running in a middleware, allows the system to execute several composed or top-level bio-inspired mechanism at the same time, all sharing the basic mechanisms implemented inside the core. Executing more than one bio-inspired mechanism at the same time is also a step forward the self-composition and self-adaptation of mechanisms, thus allowing the system to dynamically build new mechanisms as a composition of existing ones or to adapt the existing mechanisms to solve new problems.

Our long-term goal is to provide a programming framework for bio-inspired applications that abstracts away the underlying bio-inspired mechanisms driving the behaviour of the many entities composing the application. To this aim, this paper presents an execution model for such a set of core services, called BIO-CORE, which includes a core set of bio-inspired services, a core's shared data space and core interfaces, as well as a model of interaction between the applications and the core services.

This paper is structured as follows: Section 2 presents previous works existing in the literature. We then propose a Computational Model specifying the interactions between the agents participating in self-organising applications and the core itself. Section 4 presents the design of BIO-CORE. Section 5 discusses simulation results of two bio-inspired applications using BIO-CORE. Finally, we identify future work.

2 Related Works

Besides development methods focusing on iterative design and extensive simulations [Puviani et al., 2009], three main approaches for engineering self-organising applications are emphasised so far: (1) Self-organising design patterns presenting bio-inspired mechanisms according to software design patterns schemes, identifying how, when and where the patterns can be used, and providing a reusable solution for common recurrent problems; (2) Middleware that provide a support for storing, propagating and maintaining distributed tuple-based structures, facilitating design and development of bio-inspired self-organising systems by providing specific built-in features, and (3) Bio-inspired execution models for communications protocols, which provide new paradigms where self-organisation is reached by reactions among pieces of information occurring in a spontaneous way.

2.1 Self-organising Design Patterns

Self-organising mechanisms expressed as design patterns help identifying the *problems* that each mechanism can solve, the specific *solution* that it brings, the *dynamics* among the entities involved in the pattern and details of their *implementation*.

Several authors have proposed self-organising mechanisms following the software design pattern scheme [Babaoglu et al., 2006, Gardelli et al., 2007, De Wolf & Holvoet, 2007]. However, relations among the patterns are not identified, i.e. the authors do not describe how patterns can be combined to create new patterns or adapted to tackle different problems.

Focusing   on   the   relation   between   the   mechanisms, [Fernandez-Marquez et al., 2011b, Fernandez-Marquez et al., 2011a] show how the different patterns relate with each other and how some patterns are composed by others. [Fernandez-Marquez et al., 2011a] first proposed a decomposition of the Gossip mechanism into two basic mechanisms, Aggregation and Spreading, and then [Fernandez-Marquez et al., 2011b] proposed a decomposition of the Gradient Case into Spreading, Aggregation and Evaporation mechanisms, and showing how at the same time the Gradient mechanism itself is exploited by higher-level mechanisms as Morphogenesis, Quorum Sensing and Chemotaxis. The complex catalogue of self-organising design patterns is presented in [Fernandez-Marquez et al., 2012].

2.2 Middleware for Self-organising Systems

In order to facilitate the design and development of bio-inspired self-organising systems, a series of middleware proposals have been presented recently in the literature. TOTA (tuple On The Air) [Mamei & Zambonelli, 2005] provides a support for storage, propagation and maintenance of distributed tuple-based data structures. Similar approaches include MeshMDL [Herrmann, 2003] and Lime [Dept & Murphy, 2001]. Those proposals are all based on tuple space technology (i.e. shared spaces where agents indirectly exchange information), thus providing a way to implement indirect communication between agents.

While these middleware ease the task of the programmer of a self-organising system by taking care of the execution of some low-level mechanism, they still require that the programmer carefully describes and programs the mechanisms' behaviour (e.g. propagation of a gradient in a distributed system), preventing the reuse of code.

2.3   Bio-inspired Execution Models for Communications Protocols

Recent bio-inspired execution models for communications protocols provide a new paradigm, inspired by molecular processes, where self-organisation is reached by spontaneous reactions among pieces of information. The chemical metaphor was originally proposed by Gamma in 1986 as a formalism for the definition of programs without artificial sequentiality [Banâtre et al., 2001]. The basic idea underlying the formalism was to describe computation as a form of chemical reactions on a collections of individual pieces of data. Based on this chemical metaphore, we highlight two different execution models for communications protocols: (1) Fraglets [Tschudin, 2003] unify code and data into a single unit, the so called "fraglet". This single unit is a string composed of symbols values. The first symbol value is a tag that represents the instruction that is going to be executed over the fraglet. Fraglets are stored in a fraglet store in the same way as tuples are stored in a shared tuple-space; (2) Rule-based Systems [Dressler et al., 2009], where self-organisation is reached by a given set of rules acting on passive data packets. Rule-based systems follow an approach similar to the fraglets, however, a piece of data does not contain code, and a rule can operate over several pieces of data in the same execution.

The execution model proposed in this paper follows both fraglets and rule-based systems. BIO-CORE provides basic bio-inspired mechanisms under the form of "core bio-inspired services" implemented by rules. This enables to process several tuples at the same time and, thus allows complex ways of aggregation besides other operations. Moreover, pieces of data upon which services apply contain properties, similar to tags, indicating to the engine how that piece of data must be processed.

These services are intended to equip the designers and programmers of higher-level self-organising services or applications with a set of ready-to-use low-level mechanisms, whose implementation and execution is taken in charge by a specific middleware.

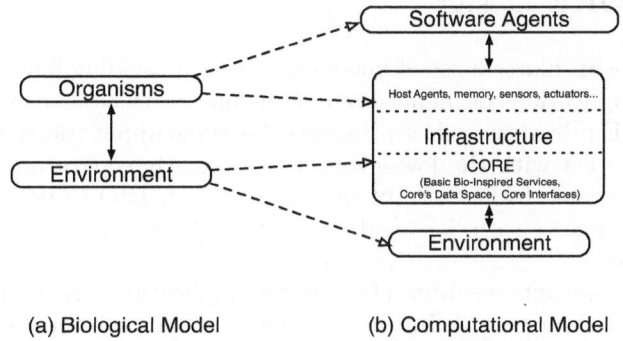

(a) Biological Model   (b) Computational Model

Fig. 1. Relevant entities of the biological and computational models

3   The Computational Model

A bio-inspired computational model, for describing the interactions be-
tween the entities participating in self-organising systems, was presented
by [Fernandez-Marquez et al., 2011a]. The computational model is as follows:
Agents are autonomous and pro-active software entities running in a Host.
The *Infrastructure* is composed of a set of connected Hosts and Infrastructural
Agents. A *Host* is an entity with computational power, communication capabilities, and may have sensors and actuators. Hosts provide services to the agents.
An *Infrastructural Agent* is an autonomous and pro-active entity, acting over
the system at the infrastructure level. Infrastructural agents may be in charge of
implementing those environmental behaviours present in nature (e.g. spreading,
aggregation and evaporation of pheromones). Finally, the *Environment* is the
physical space where the infrastructure is located.

In this paper we extend the computational model presented
in [Fernandez-Marquez et al., 2011a]. The new model, showed in Figure 1,
adds the notion of BIO-CORE, which provides basic bio-inspired mechanisms,
ready-to-use as "first-class" entities by higher-level services or applications
(simply called CORE in Figure 1. BIO-CORE aims at decoupling the agents
from the environment's behaviour by providing a *virtual environment* (as
opposed to the actual real-world environment) where more than one bio-
inspired algorithm can be executed at the same time, reusing implementation
and enabling the creation of new bio-inspired mechanisms for solving new
problems or dynamically adapting existing ones. BIO-CORE is composed of: 1.
a Core's Data Space, where agents deposit and retrieve data; 2. a set of Basic
Bio-Inspired Services implementing basic bio-inspired mechanisms through rules
applying on data deposited in the data space; and 3. Core Interfaces providing
primitives for the agents to interact with BIO-CORE, for accessing other cores
in neighbouring nodes, and for accessing sensors and actuators of the local
node. BIO-CORE is embedded into each device participating in the system.

4   BIO-CORE Design

BIO-CORE encapsulates a set of low-level services providing bio-inspired mechanisms that applications or higher-level bio-inspired services can exploit and rely on. BIO-CORE provides a set of primitives for these applications and high-level services to interact with the low-level services, clearly separating the responsibilities of the agents from those of the environment. BIO-CORE also provides a shared data space for services and applications to exchange data and interact with each other.

BIO-CORE advantages are: (1) Several applications or higher-level bio-inspired services can be running in the same virtual environment re-using the services provided by BIO-CORE (i.e. reusing code); (2) It makes it easier to model and implement bio-inspired applications, since agents' behaviour is decoupled from the environment, and the low-level services provided by BIO-CORE are still running in the middleware, ready to be executed on demand; (3) Since several bio-inspired mechanisms can be running at the same time, BIO-CORE is a first step towards self-composition of mechanisms.

The basic bio-inspired services provided by BIO-CORE are those identified during our work on defining design patterns for bio-inspired mechanisms, where we expressed high-level bio-inspired mechanisms as a composition of lower-level ones [Fernandez-Marquez et al., 2011b, Fernandez-Marquez et al., 2011a]. Namely, the mechanisms provided by BIO-CORE are: Spreading, Evaporation, Aggregation and Gradients. These mechanisms present common characteristics: (1) in biological processes they are mainly executed by the environment; (2) they occur both in macro- and micro-level systems (e.g. spreading mechanism can be found in ants colonies coordination or in signaling pathways between cells); and (3) they are at the basis of more complex self-organising mechanisms (e.g. gossip is a combination of aggregation and spreading).

Figure 2(a) shows the interactions between agents belonging to applications or to high-level bio-inspired services and BIO-CORE, which contains low-level services providing basic bio-inspired mechanisms. Figure 2(b) describes the relations between BIO-CORE and agents belonging to applications inside a given host. Agents have access to the *Communication Device, Sensors and Actuators* provided by the Host. Agents communicate with BIO-CORE using the *Agent Interface*. This interface is directly connected to the *Core's Data Space*, a shared data space allowing agents to deposit and retrieve data through specific primitives. The data deposited into the Core's Data Space is processed by the *Core Engine*. Namely, the Core Engine is composed of an Infrastructural Agent (IA), and a set of rules implementing the core services. The IA has access to the Communication Device and Sensors/Actuators provided by the Host through the *Communication Interface* and *Sensors/Actuators Interface* respectively. It allows data from the Core's Data Space to be sent to neighbouring Core's Data Spaces, to provide access to Sensor's data by inserting Sensor reads into the Core's Data Space, and to instruct specific Host's actuators to perform some action (e.g. move the wheels of a robot in a certain direction).

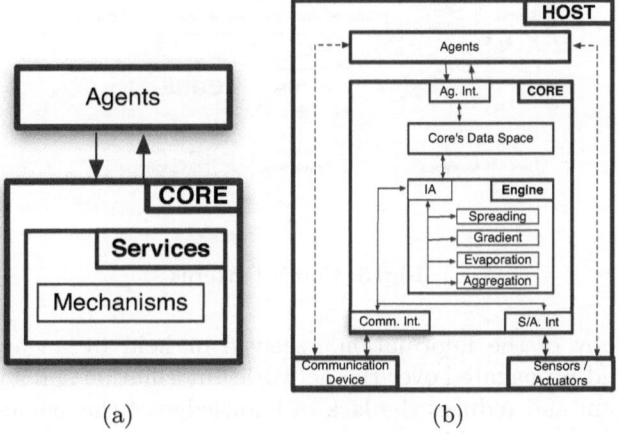

Fig. 2. System's Architecture

4.1 BIO-CORE Engine

The BIO-CORE engine is composed of the Infrastructural Agent (IA) and a set of rules that implement the low-level services offered by BIO-CORE. Basically, the IA is responsible for applying the rules to the set of data stored in the Core's Data Space according to the data's properties. The IA is also responsible for managing the internal interfaces (e.g. sending or receiving data to other Core's Data Spaces or acting over the sensor or actuators).

The rules are transitions that provide BIO-CORE with chemical reactions similar to chemical machine models [Banâtre et al., 2001, Dressler et al., 2009] and make the IAs act over the Core's Data Space in a completely distributed and decentralised way. Rules provide a simple way to define the environment's behaviour, emulating the laws of nature, and providing the environment with an autonomous and proactive behaviour (i.e. Applications do not actually call those rules, rules are applied dynamically when necessary). Indeed, the behaviour of BIO-CORE over the data stored in the Core's Data Space depends on the data's properties, in the same way as in the real word, the environment acts over the entities depending on their properties and the nature's laws (e.g. gravity, diffusion, aggregation, etc.).

Figure 3 shows these basic mechanisms provided by BIO-CORE, and how high-level bio-inspired mechanisms are composed from these core mechanisms.

A full description, under the form of design patterns, can be found in [Fernandez-Marquez et al., 2011b, Fernandez-Marquez et al., 2011a].

Spreading Pattern. The Spreading Pattern [Fernandez-Marquez et al., 2011a] is a basic pattern for information diffusion/dissemination. The Spreading Pattern progressively sends information over the system using direct communication among agents, allowing the agents to increment the global knowledge of the system by using only local interactions.

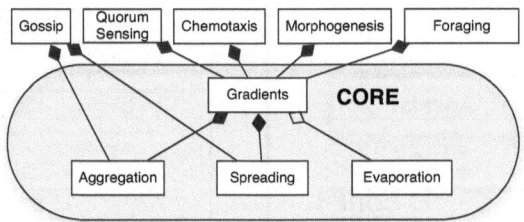

Fig. 3. Core's Patterns

Rule: A copy of the information, received or held by an agent, is sent to neighbours and propagated over the network. Information spreads progressively over the system and reduces the lack of knowledge of the agents while keeping the constraint of the local interaction.

The Spreading Pattern is one of the most used in the literature, and it appears in important higher-level bio-inspired patterns, such as, Morphogenesis Pattern, Quorum Sensing Pattern, Chemotaxis Pattern, Gossip Pattern, and Gradient Pattern [Fernandez-Marquez et al., 2012].

Aggregation Pattern. The Aggregation Pattern [Gardelli et al., 2007], is a basic pattern for information fusion. The dissemination of information in large-scale systems deposited by the agents or taken from the environment may produce network and memory overload, thus, the necessity of synthesising the information. The Aggregation Pattern reduces the amount of information in the system and assesses meaningful information.

Rule: Aggregation consists in locally applying a fusion operator to process the information and to synthesise macro information. This operator can take many forms, such as filtering, merging, aggregating, or transforming. In BIO-CORE, the aggregation service fuses the information present in the Core's Data Space. Information comes from the real-world environment (through sensors reads like temperature, humidity, etc.), from other agents (i.e. through communication with other core's space) or from agents running in the Host interacting directly with the Core's Data Space. The aggregation process terminates when aggregation leads (through one or more applications of the aggregation law) to an atomic information.

The Aggregation Pattern used in conjunction with the Evaporation and Spreading Patterns is at the basis of the digital pheromone and thus the Foraging Pattern.

In BIO-CORE enabling simultaneously Aggregation, Evaporation and Spreading services on a piece of data allows: to create digital pheromones; to perform Gossip; and to run applications that exploit gradients.

Evaporation Pattern. Evaporation is a pattern that helps to deal with dynamic environments where information used by agents can become outdated. In real world scenarios, the information changes with time and its detection, prediction, or removal is usually costly or even impossible. Thus, when agents have to adapt their behaviour according to information from the environment, information gathered recently must be more relevant than information gathered a long time ago.

Rule: Evaporation is a mechanism that progressively reduces the relevance of information.

In BIO-CORE, enabling the Evaporation and Gradient services allows to build dynamic gradients, making them adaptable to topology changes. The Evaporation service used in conjunction with the Spreading and Aggregation services allows to create digital pheromones. Used on its own, the Evaporation service allows the data deposited in the Core's Data Space, and to which it applies, to become outdated and to be removed by the IA.

Gradient Pattern. The Gradient Pattern focuses on large scale system, where agents suffer from lack of global knowledge to estimate the consequences of their actions or the actions performed by other agents beyond their communication range. Using the Gradient Pattern, information is spread from the location where it was initially deposited and it is aggregated when it meets other information. Thus, agents that receive gradients have information that come from beyond their communication range, increasing the knowledge of the global system not only with gradient's information but also with the direction and distance of the information source.

Rule: The Gradient Pattern is an extension of the Spreading Pattern where the information is propagated in such a way that it provides an additional information about the sender's distance and direction. Additionally, the Gradient Pattern uses the Aggregation Pattern to merge different gradients created by different agents or to merge gradients coming from the same agent but through different paths.

4.2   BIO-CORE Data

We could envisage different ways of dynamically applying services on data. Here, we consider the use of *properties* attached to data. Indeed, data are passive entities that, once deposited into the Core's Data Space, are subject to the actions of BIO-CORE services (e.g. modified, cloned or removed) depending on their properties. Services provided by BIO-CORE are then activated on demand by the applications or higher-level services by modifying the data's properties. The actual activation occurs through the Infrastructural Agent that takes care of identifying the appropriate service. The interactions between the Infrastructural Agent and the data are defined by the set of low level services performed in the BIO-CORE Engine.

Data properties are defined in table 1. Basically the idea is that if a piece of data has the Evaporate property set to true, then the Evaporation service will be executed over the data. A piece of data can have several properties equals to true at the same time, thus enabling multiple services to act over it (e.g. data that is spread, aggregated, evaporated and subject to gradient could be a

Table 1. BIO-CORE Data's Properties

Property	Description
ID	Unique identifier
Evaporate	Activate the Evaporation service
Aggregate	Activate the Aggregation service
Spread	Activate the Spreading service
Gradient	Activate the Gradient service. Automatically this property also enables the spread and aggregation properties.
Information	Actual information stored in the data

digital pheromone; data that is spread and aggregated may be the subject of gossip mechanism; data that is subject to gradient and evaporated can be used to create dynamic gradients, etc. . .). Moreover, data contains also parameters for defining the probability of evaporation, the kind of evaporation, the kind and frequency of aggregation, etc.

4.3   BIO-CORE Interfaces

BIO-CORE defines three different interfaces: (1) an external interface for the Agents to communicate with the Core's Data Space. Agents adapt their behaviour according to the data retrieved from the Core's Data Space and conversely, by inserting appropriate data into the Core's Data Space delegate environmental responsibilities, such as, spreading, aggregation or evaporation to BIO-CORE; (2) an internal interface for exchanging data among Core's Data Space of different Hosts; and (3) an internal interface for the Core's Data Space to communicate with Sensors and Actuators of the Host.

Data is exchanged between the Core's Data Space and the agents, and vice-versa, through external interface primitives for creating, depositing and retrieving data.

5   Simulation Results

This section shows the design and development of two bio-inspired applications using BIO-CORE. The first application, "Reaching an agreement" is based on the Gossip mechanism. The real-world environment consists of a set of Hosts (or nodes) each equipped with an initial random colour. The goal is to reach an agreement on the colour of the nodes, where all nodes share the same colour.

The second application is "Regional Leaders Election", where Spreading and Evaporation services are used to assign the leader and member roles to the nodes participating in the systems. The goal of this section is to analyse the design and feasibility of these applications using BIO-CORE.

5.1   Reaching an Agreement

As it was presented in [Fernandez-Marquez et al., 2011a], Gossip is a composed mechanism, where the Spreading and the Aggregation mechanisms are used

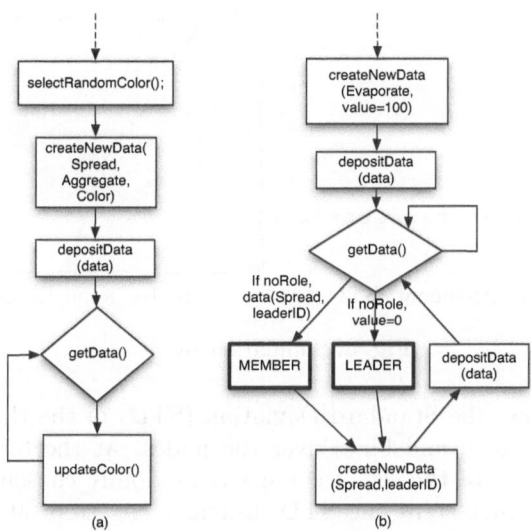

Fig. 4. (a) Reaching an agreement (b) Regional Leaders Election

simultaneously. In the gossip process the information is sent over the network using the Spreading mechanism and it is aggregated at each node with the local information by using the Aggregation mechanism.

In this simulation the system has to reach an agreement on the nodes' colour. Initially, each node has a random colour and during the simulation the colour of the nodes must converge to the same colour. Moreover, if new nodes appear the system must be able to deal with it and reach the agreement again taking into account new colours: if the new nodes are blue, the whole system will change the colour towards a blue shade. The aggregation operator used in the simulation is the average between the three components of colour (i.e. red, green and blue, where each component is in [0-255]). This simulation executed with 500 nodes randomly placed in a 1200x700 meters bi-dimensional space, the mobility pattern used is random walk and the communication range for each node is 60 meters.

Figure 4(a) shows the flow diagram for each agent in the "Reaching an agreement" application. The different steps are presented as follows: (1) each agent in each node initially chooses a colour at random; (2) it creates a Core data, with properties Spread, Aggregate set to true and the colour as Information; (3) the data is deposited into the Core's Data Space; (4) Spreading and Aggregation services of BIO-CORE then act on these data: all nodes spread the data, on each node aggregation then acts on all data whose Aggregate property is set to true, averaging the colour Information (only one piece of data per node will remain, containing the average colour of the node and that of its neighbours); (5) periodically agents check their local Core's Data Space, and retrieve new aggregated core data, (6) each agent then updates its colour to the one contained in the Information field provided by the retrieved Core data and returns to step 5.

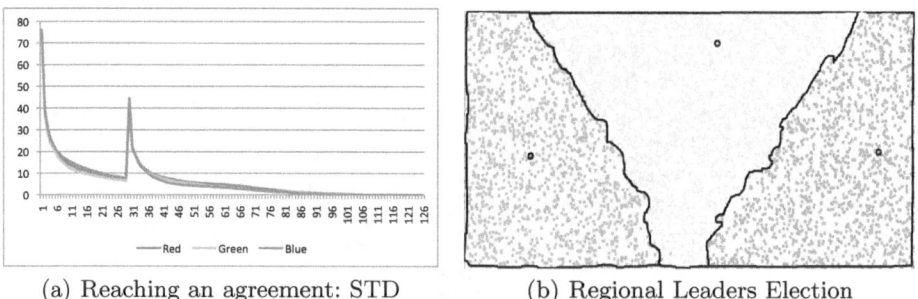

(a) Reaching an agreement: STD   (b) Regional Leaders Election

Fig. 5. Simulation results

Figure 5(a) shows the Standard Deviation (STD) of the three colour's components (i.e. red, green and blue) over the nodes. At the beginning the STD is maximum because of the set of colours is randomly chosen. The simulation shows how in the first steps the STD decreases. At step 30, new nodes with random colours are added into the system. These new nodes increment the STD but the system easily overcomes this environmental change and reaches the final agreement after few steps, reaching an STD equals to 0 (i.e. all the nodes share the same colour). It is out of the scope of this work to improve the performance of the gossip algorithm. We are interested here in showing how basic bio-inspired services apply on data provided by applications or higher-level services.

5.2   Regional Leaders Election

The regional leaders election was presented as an application example exploiting amorphous computing primitives in [Abelson et al., 2000]. The goal is to split the network into disjoint groups each led by one node, following a decentralised and distributed process. Initially nodes have no identified role. Once the system converges, the network is broken up into contiguous domains each composed of one leader and members.

Figure 4(b) shows the flow diagram of the agents behaviour in the Regional Leaders Election process using BIO-CORE. The steps are as follows: (1) each agent creates a Core data with properties Evaporate set to true and Information equals to 100 representing the relevance value; (2) each agent deposits this data into the Core's Data Space; (3) BIO-CORE periodically decreases the relevance value; (4) if an agent has no assigned role yet, it checks the local Core's Data Space periodically. If the Information (relevance value) of the Core data has reached 0 the agent decides to become a leader, it then creates and deposits a new Core data in the Core's Data Space with properties Spread set to true, including its agentId in the ID properties. BIO-CORE will then spread this new Core data. If the Information Core data is not yet 0, but another Core Data with information (Spread, agentId) is found in the local Core's space, the agent decides to become a member of the leader whose id is equal to agentId.

Both applications implemented using BIO-CORE have reached the desired emergent behaviour. The flow diagrams show that both design have been reduced from the original ones, in the sense that most of the responsibilities are performed by BIO-CORE. Moreover, the BIO-CORE services have been reused by two applications making easier not only the design phase, but also the implementation. It exists a wide number of applications based on the basic mechanisms implemented inside the BIO-CORE. Thus, BIO-CORE is a first step to create systems where bio-inspired applications can be executed, reusing code, sharing the same virtual environment as biological process share different mechanisms in a real environment.

6   Conclusions

This paper presents BIO-CORE, an execution model for a set of low-level services providing basic bio-inspired mechanisms that applications or high-level bio-inspired services can exploit and rely on. BIO-CORE provides primitives for those applications interacting with the low-level services, clearly separating the responsibilities of the agents from those of the environment. BIO-CORE design is presented based on shared data space technology and rules. This is one way of considering the implementation of BIO-CORE, other techniques or existing middleware could be used. This paper focuses on engineering bio-inspired self-organising systems, providing a core for designing and implementing bio-inspired applications. BIO-CORE feasibility is analysed using two different applications, "Reaching an Agreement" and "Regional Leaders Election", where both simulations have reached the desired emergent behaviour.

Future works will focus on three different directions: (1) extending the catalogue of mechanisms and the relations between them, in order to extend BIO-CORE giving support for the maximum number of bio-inspired applications; (2) implementing BIO-CORE for Android OS, analysing different implementations, and (3) even when the services' parameters can be set up by passing arguments down from application to the BIO-CORE, we plan to work on self-composition of services and self-adaptation of parameters in order to avoid a complex parameterisation of the services and to provide a better performance against environmental changes.

Acknowledgments. This work has been supported by the EU-FP7-FET Proactive project SAPERE Self-aware Pervasive Service Ecosystems, under contract no.256873.

References

[Abelson et al., 2000] Abelson, H., Allen, D., Coore, D., Hanson, C., Homsy, G., Knight Jr., T.F., Nagpal, R., Rauch, E., Sussman, G.J., Weiss, R.: Amorphous computing. Commun. ACM 43(5), 74–82 (2000)

[Babaoglu et al., 2006] Babaoglu, O., Canright, G., Deutsch, A., Caro, G.A.D., Ducatelle, F., Gambardella, L.M., Ganguly, N., Jelasity, M., Montemanni, R., Montresor, A., Urnes, T.: Design patterns from biology for distributed computing. ACM Trans. on Autonomous and Adaptive Sys. 1, 26–66 (2006)

[Banâtre et al., 2001] Banâtre, J.-P., Fradet, P., Le Métayer, D.: Gamma and the Chemical Reaction Model: Fifteen Years After. In: Calude, C.S., Pun, G., Rozenberg, G., Salomaa, A. (eds.) Multiset Processing. LNCS, vol. 2235, pp. 17–44. Springer, Heidelberg (2001)

[De Wolf & Holvoet, 2007] De Wolf, T., Holvoet, T.: Design Patterns for Decentralised Coordination in Self-organising Emergent Systems. In: Brueckner, S.A., Hassas, S., Jelasity, M., Yamins, D. (eds.) ESOA 2006. LNCS (LNAI), vol. 4335, pp. 28–49. Springer, Heidelberg (2007)

[Dept & Murphy, 2001] Dept, A.M., Murphy, A.L.: LIME: A Middleware for Physical and Logical Mobility. In: Proc. of the 21st Int. Conf. on Distributed Computing Systems, ICDCS 2001, pp. 524–533. IEEE Computer Society (2001)

[Dressler et al., 2009] Dressler, F., Dietrich, I., German, R., Krüger, B.: A rule-based system for programming self-organized sensor and actor networks. Comput. Netw. 53, 1737–1750 (2009)

[Fernandez-Marquez et al., 2011a] Fernandez-Marquez, J.L., Arcos, J.L., Di Marzo Serugendo, G., Casadei, M.: Description and Composition of Bio-Insp. Design Patterns: the Gossip Case. In: Int. Conf. on Engineering of Autonomic and Autonomous Syst. (EASE), pp. 87–96. IEEE Computer Society (2011a)

[Fernandez-Marquez et al., 2011b] Fernandez-Marquez, J.L., Arcos, J.L., Di Marzo Serugendo, G., Viroli, M., Montagna, S.: Description and Composition of Bio-Inspired Design Patterns: The Gradient Case. In: Workshop on Bio-Insp. and Self-*Algorithms for Distributed Systems (BADS), pp. 25–32. ACM (2011b)

[Fernandez-Marquez et al., 2012] Fernandez-Marquez, J.L., Di Marzo Serugendo, G., Montagna, S., Viroli, M., Arcos, J.L.: Description and Composition of Bio-Inspired Design Patterns: a complete overview. Natural Computing Journal (invited paper, submitted, 2012)

[Gardelli et al., 2007] Gardelli, L., Viroli, M., Omicini, A.: Design Patterns for Self-Organizing Multiagent Systems. In: De Wolf, T., Saffre, F., Anthony, R. (eds.) 2nd International Workshop on Engineering Emergence in Decentralised Autonomic System (EEDAS), pp. 62–71. CMS Press (2007)

[Herrmann, 2003] Herrmann, K.: MESH Mdl " A Middleware for Self-Organization in Ad Hoc Networks. In: Proc. of the 23rd Int. Conf. on Distributed Computing Systems, ICDCSW 2003. IEEE Computer Society (2003)

[Mamei & Zambonelli, 2005] Mamei, M., Zambonelli, F.: Programming stigmergic coordination with the TOTA middleware. In: Proc. of the 4th Int. Joint Conf. on Autonomous Agents and Multiagent Systems, AAMAS, pp. 415–422. ACM (2005)

[Puviani et al., 2009] Puviani, M., Di Marzo Serugendo, G., Frei, R., Cabri, G.: Methodologies for Self-Organising Systems: A SPEM Approach. In: Proc. of the 2009 IEEE/WIC/ACM Int. Joint Conf. on Web Intelligence and Intelligent Agent Technology, WI-IAT, pp. 66–69. IEEE Computer Society (2009)

[Tschudin, 2003] Tschudin, C.F.: Fraglets - a Metabolistic Execution Model for Communication Protocols. In: In Proceeding of 2nd Annual Symposium on Autonomous Intelligent Networks and Systems (AINS), Menlo Park (2003)

Comparison of Ant-Inspired Gatherer Allocation Approaches Using Memristor-Based Environmental Models

Ella Gale, Ben de Lacy Costello, and Andrew Adamatzky

Unconventional Computing Group, University of the West of England, Bristol, UK
ella.gale@uwe.ac.uk
http://uncomp.uwe.ac.uk/index.html

Abstract. Memristors are used to compare three gathering techniques in an already-mapped environment where resource locations are known. The All Site model, which apportions gatherers based on the modeled memristance of that path, proves to be good at increasing overall efficiency and decreasing time to fully deplete an environment, however it only works well when the resources are of similar quality. The Leafcutter method, based on Leafcutter ant behaviour, assigns all gatherers first to the best resource, and once depleted, uses the All Site model to spread them out amongst the rest. The Leafcutter model is better at increasing resource influx in the short-term and vastly out-performs the All Site model in a more varied environments. It is demonstrated that memristor based abstractions of gatherer models provide potential methods for both the comparison and implementation of agent controls.

Keywords: memristor memristors networks gathering model multi-agent systems ants Leafcutter Atta gatherer animal behaviour.

1 Introduction

1.1 Memristors

Memristors are an emerging technology with anticipated wide-spread applications in neuromorphic computing, artificial intelligence and green technology. The memristor is the 4th fundamental circuit element predicted to exist in 1971 [1] and was only brought to wide-spread attention in 2008 [2]. A memristor differs from the resistor by being able to store a state ie. it possesses a memory. Standard modern-day computers separate the processor and memory to different physical places, whereas the memristor can be used as both. As this is similar to how the brain works [3], it is thought that artificial intelligences and A.I.-like systems would be easier to create with a memristor-based system [4]. Memristor-based systems have other advantages over transistor-based systems: their memory is not volatile, and thus removal of the power source does not erase data [5]. From a green computing point of view memristors only draw

E. Hart et al. (Eds.): BIONETICS 2011, LNICST 103, pp. 73–84, 2012.

power when accessed. Again, this is similar to the brain, and fits well with neuromorphic computing paradigms where we are concerned with the propagation of spiking direct current signals rather than repeated A.C. clock cycles.

Thus far, we [6,7] and others [8] have focused on utilising memristors in spiking networks, using a bottom up approach to building memristor computers. Such networks are often considered in terms of their use as control systems for autonomous or distributed systems and accordingly we focused on path-finding by a single autonomous agent. While the bottom-up approach is good for increasing complexity of a single agent, we now present the complimentary approach which has uses for a different class of problems.

Memristors can be used at a higher level of abstraction such as in a top-down approach to model the environment. An example of this is maze-solving where the maze is modeled as a grid of memristors [9] (note path-finding via grids of resistors had been done prior to this [11,10].). By connecting the source and voltage to the entrance and exit of the maze, the resultant voltage drop across each memristor in the solution paths causes its resistance to change. This allows the memristors to 'solve' the maze in the time taken for one memristor to fully switch, regardless of the size of the maze or the length of the solution path. Thus far, this method has only been simulated, but the real interest lies in using the memristors in hardware to allow the laws of physics to solve the problem in a much shorter time.

Gathering as a process can be conceptually separated into two parts: firstly finding the resource (resource location and path finding) and secondly harvesting or gathering the resource (following laid-down paths and returning with resources). Here we concentrate here on the problem of efficient gathering in an environment in which the resources have been located. The gatherers under study could be ants gathering food or autonomous agents mining useful resources for a factory or collecting samples for scientific research.

1.2 Ants

The existence of pheromones and their use in guiding ants was first reported in 1880 [12] and the positive reinforcement of trails laid by ants was reported in 1962 [13]. Since then the picture has become much more detailed: there are several pheromones [14] which last for different times [15] with some that reverse previous instructions [16] that together tell the ants where to go and there are additional geometrical cues which tell them where they are [18]. Nonetheless, the simple model of ants exploring their environment and responding by positively reinforcing pathways has given rise to several different ant optimisation algorithms from the original [19], to Ant Colony Optimisation for optimisation problems [20,21] and extensions of that to the Ant Colony System [22] and Rank based Ant system [23]. These algorithms have been successfully applied to various problems such as the traveling salesman [25] and other np-hard problems [24].

Given that study of ant searching behaviour has given rise to such a range of useful computational techniques, we chose to ask what they can teach us about harvesting. Different ant species have different behaviours, but in general they seek to maximise the colony's energy intake [26], usually focusing on energy maximisation rather than harvesting time minimisation. For example, *Solenopsis germinata* will preferentially go to the closer, bigger and higher concentration food sources over the more distant, smaller sources [27]. The Leafcutter ants, *Atta*, have a slightly different technique, they preferentially go to only the best leaf site when they are in a resource rich environment and spread out to a greater diversity of trees/sites when in a resource poor environment[28].

Some ants separate the finding of resources from their gathering. For example, once seed-gathering ants have found a food source, they will return to search that area, completely ignoring any seeds placed near their path [29]. Similarly, Leafcutter ants will create trails which the gatherer tend to stick to. Thus, it is valid to concentrate on the behaviour of gatherers once the environment has been explored.

The dynamics of ant colonies are non-linear, which makes memristors, as non-linear devices, well suited to model them. To demonstrate the usefulness of memristors as an environmental model, we look at the problem of the most efficient gathering techniques in two very different situations, that of a resource rich and resource poor environments. We will focus on three different models of behaviour: 1, the Sequential Gathering technique where all the gatherers go to each food source in turn, starting with the best; 2, the All Sites model where the gatherers are split between all the food sources; and 3, the Leafcutter model where all the ants deplete the best resource and then spread out amongst the rest.

2 The Memristor Model of the Environment

2.1 The Model Memristor

Memristance M, relates charge, q and magnetic flux φ by $\varphi = M(q)q$. As M is a function of q, the memristance is controlled by the amount of charge on the memristor, this is related to the current that has passed through it, and so q the memory of the device. q is a function of time, t, and at any instant in time the memristor acts like an Ohmic resistor in that $V = RI$, where R is the instantaneous resistance, and memristance is the time-varying resistance.

Memristors have physical limits to the possible resistance values: the lower limit R_{on} and the upper limit R_{off}. The first model of memristance to relate it to physical measureables [2] gave the memristance as

$$M(q) = R_{\text{off}} - R_{\text{off}}R_{\text{on}}\beta q, \qquad (1)$$

where β is $\frac{\mu_v}{D^2}$. This model refers to the Strukov memristor [2] where two different forms of titanium dioxide (with differing resistivities) are interconverted based on the drift of oxygen vacancies, as governed by the oxygen vacancy ion mobility, μ_v, across a thin-film of thickness D. However, for our current model, we can view β as a parameter which varies based on the material properties of the device (for a discussion on the effect of varying this parameter, see [7]).

2.2   Methodology

Using Memristors to Model the Environment. The test environment is five different food sources different distances from the nest which is modeled by five memristors in series with a 5V potential difference applied across them. The total voltage drop does not change and relates to the number of available gatherers in the system. The memristors are set to the ON state (the low resistance mode) so that an increase in charge will eventually switch the device into the OFF state. Thus, for ease of use, the memristance is modeled as $M(q) = R_{on} + R_{off} R_{on} \beta q$, which is equivalent to the equation above if we include the boundary condition: $R_{on} \leq M(q) \leq R_{off}$.

At the start of the simulation, each memristor is charged to a different degree, ie $M(t = 0)$ varies, and this number encompasses the amount of resource, and the length and difficulty of the path (the more resources present and the easier the path, the lower the value of $M(t = 0)$). This is done by changing R_{on} in the model, in the lab this would be done by charging the memristor up by a set amount before the start of the experiment. β is set to 1 in this model for all memristors, and it represents the difficulty getting resources from the site. The memristor equation gives a curve, and β controls the curvature [7] and due to the non-linearity of the curve this includes the law of diminishing returns whereby, because gatherers get in each other's way, each additional gatherer at a resource gives a smaller productivity gain as the number of gatherers at a resource rises. M increases as the food is removed, until it hits R_{off}, at which point the resource is considered depleted (the amount of resources at the start can vary, but when depleted, it is depleted no matter how much there was to begin with). As M has a hard top limit, the resources are scaled to start in different places. The sum of the current is the rate of resource influx at the nest. In this situation, a real memristor system would not have a limited current, but the total amount of resources in a test environment should be the same, thus we use

$$\% \text{ of gathered resource on step } n = \sum_{t=0}^{t=n} \frac{I(t)}{I(D)} \qquad (2)$$

where $I(D)$ is the current on the step where the environment is entirely depleted. Table 1 summarises the relationships between the model memristor circuit and the ant model.

For the rich environment, the memristors are set to: $M_1(0) = 1\Omega$, $M_2(0) = 2\Omega$, $M_3(0) = 0.5\Omega$, $M_4(0) = 15\Omega$ and $M_5(0) = 4\Omega$. The poor environment contains one very good site $M_1(0) = 0.5\Omega$, and the rest are set to 60Ω, 70Ω, 80Ω and 90Ω respectively (the order does not matter). Although these are given in terms of Ω for convenience, because β is a changeable parameter in this model these are really reduced units.

Using Memristors to Model Different Gathering Techniques. We simulated a series circuit of 5 memristors to model the ants going to all the sites, and this is the All Sites model. The conceptually simplest gathering technique, the

Table 1. The equivalence between the memristor circuit's physical properties and the model of an ant colony gathering in a known environment

Memristor Physical property	Ant colony model
Memristor	Food site
Total voltage drop	Total number of ants
Voltage drop across an individual memristor	Number of ants at that food site
R_{off}	Food site depleted
R_{on}	Measure of the Amount of food present, distance to food source and quality of food. at a food site
β	Measure of the difficulty of extracting food over time
Total current	Rate of food influx to the nest.
Cumulative sum of current	food influx at the nest.

Sequential model, is to send all the ants to each food site in decreasing order of their richness. To calculate this, each of the memristors was charged up individually until it reached R_{off}, whereupon the next memristor was charged. To make it a valid comparison, the single memristors were left drawing current once they had reached R_{off} as this could happen in the other two models (the value of this current draw is small). The Leafcutter model is related to the All Sites model: the best memristor is run singly, then when depleted, it's left drawing a minimal current whilst the All sites model is applied to the remaining 4 memristors.

3 Results

3.1 Example of How the Memristance and Voltage Operates in the All Site Modeled Electronic System

This discussion uses a graph taken from the All Site model in the rich environment simulation, the same principles apply in the poor environment simulation. The memristance changes as shown in figure 1, the memristance starts at R_{on} and changes as a function of the integral of the current passed through the system until it reaches R_{off}. In terms of the gathering model, the lower R_{on}, the more resources available at that position initially. As β is the same for each memristor in this example, the curvature of each memristor is the same and hence the rate of increase in difficulty in resource gathering is the same across all the sites. For this reason, the resistance change does not effect the voltage

78   E. Gale, B. de Lacy Costello, and A. Adamatzky

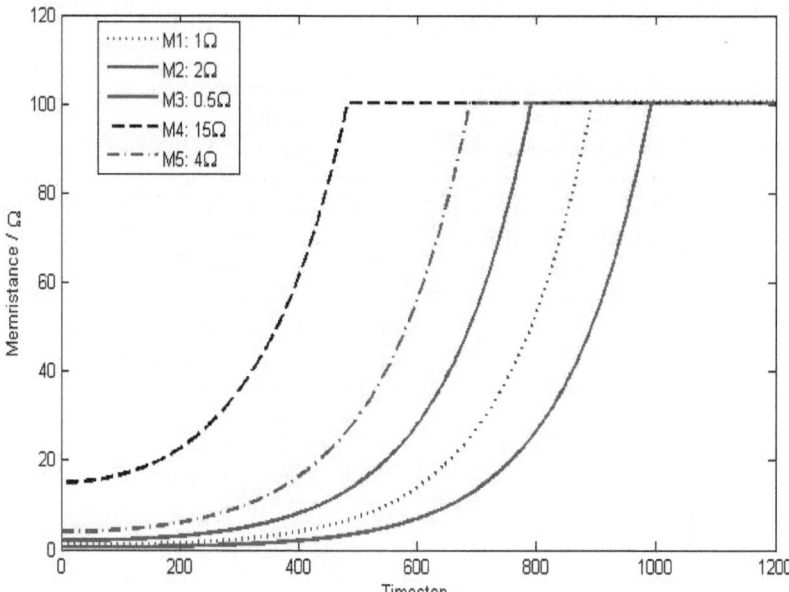

Fig. 1. How the memristance changes over time / how the food sites are depleted. Each food site starts with a different amount of food. The lower the number, the more total resource in that position. All resources are used up when they are equal to R_{off} which is set to 100. The starting values are 1,2,0.5,15 and 4 for the five memristors.

under 500 steps (see figure 2), because the voltage is shared between memristors in proportion to their relative resistances.

Figure 2 shows how the voltage drop across each memristor changes with time step. Once the 4th food source has been depleted, its share of the workers drops drastically, with most going to the next closest to being depleted, site 5, and when this is depleted, most workers move to the next resource. This continues with the number of workers assigned rising on the remaining resources as others are depleted, until time step 992 when all the resources are depleted.

4   Comparison between Gathering Approaches

4.1   Resource Rich Environment

As the maximum value of M is 100Ω, even the highest memristor starting value, 15Ω, still represents a good resource site. The rate of resource influx at the nest is the normalised cumulative current which is plotted in figure 3.

The worst model, in our opinion, is the Sequential model: it takes longer than the All Sites model, and although it beats it for resource influx in the short-term this advantage is lost after a very short time. It depletes the total environment quicker than the Leafcutter allocation method, but does not have that model's advantage of a high rate of initial influx.

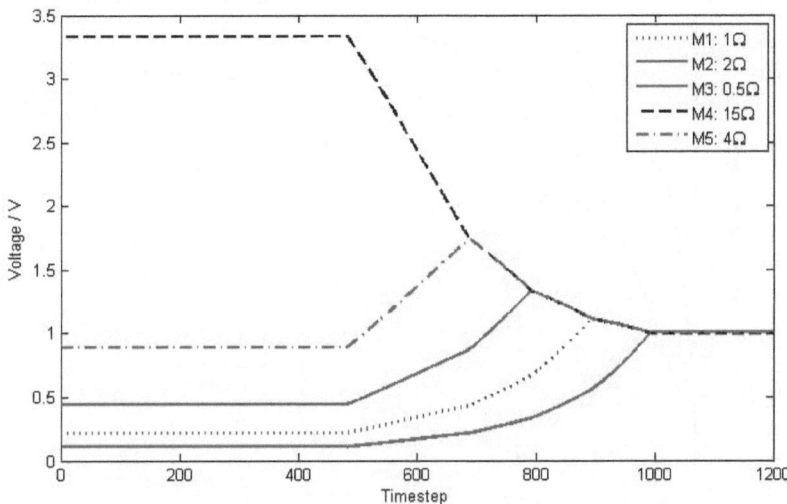

Fig. 2. The voltage across each memristor as a function of time. These curves show how ants are spread between the food sites. All sites are depleted when the voltage drop across all the memristors is equal.

To answer which is the best gathering method, we first need to ask what we expect from the best method, the quickest influx of resources or the shortest time to completely deplete the surrounding environment. The Leaf Cutter model has a clear advantage at short-times because the gatherers are concentrating on the best resource in the system. This has a cost: the Leaf-cutter model takes much longer than the other two, 2978 time steps, compared to 967 for the All Sites model and 1274 for the Sequential model, which is 108% longer than the shortest time. There is a finite amount of resources, but due to the diminishing returns, it can take different amounts of time to get them all and the measure of this difference is the differing efficiencies of the gathering techniques. In terms of fully depleting the environment, the Leaf Cutter allocation method is less efficient in this environment.

This fits perfectly with the entomological observation that ants aim to optimise the maximum energy/resource intake rather than minimize the time taken. Also, the Leafcutter method is based on the ant's strategy when in rich environments, so we might expect this model to fare well in this simulation. In the wild, there are advantages to gathering the best resource first, it prevents competitors from taking it and mitigates against any change in the environment.

However, in a rich environment where the gatherers can expect to be relatively undisturbed by competitors the All Sites model is the better approach as it allows the gatherers to entirely deplete the environment in under half as much time. Although the Leafcutter model is quicker at the start (taking 44 steps to gather 50%), the All Sites is still fast having gathered 50% of the total resources in under a 5th (17%, 173 timesteps) of the total time. Similarly, although the Sequential

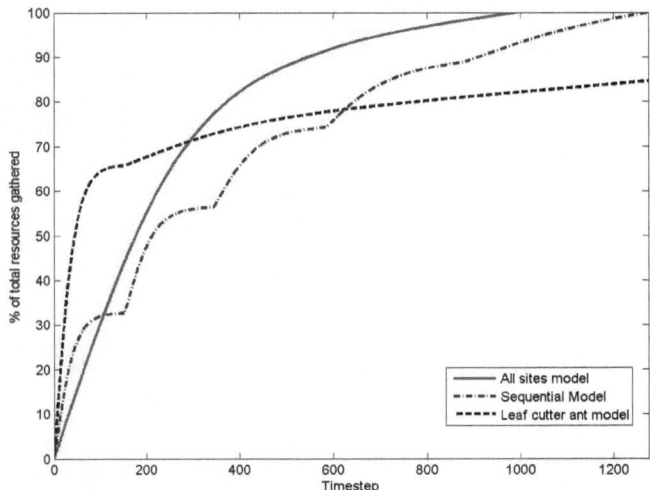

Fig. 3. Comparison of three gathering models. The Leaf cutter model is better in the short terms but takes 2978 time steps to reach 100% (all resource consumed), this is not shown in this figure.).

model is quicker than the All Sites at the start, the productivity gains trail off quickly (over 200 time steps).

Both the Leafcutter model and the Sequential model start with the best resource site first. This may seem like an obviously good idea, and interestingly, it is not how the All Sites model works. Instead the majority of gatherers go to the worst resource site first! This is because the difficulty of gathering increases with time as the 'low-hanging fruit' is taken. Thus, even the worst site yields its best resource output to begin with, and the distance to it is compensated by the high productivity at the other sites.

Once the worst resource site is depleted, that frees up workers to go to the other sites and compensate for decreasing productivity at those. Thus, if the resources are in an environment are known, it is more efficient to spread agents out amongst all of them, (with more to the worse resources) than to deplete each resource in turn. Even depleting the best resource first will delay the time taken to gather from the rest due to diminishing returns. Note, if it is desirable to try to regulate resource influx the All Sites model would be the best choice.

4.2   Resource Poor Environment

In this environment, the All Sites model is not very good, as it drastically slows the time taken to deplete the good resource and the environment (both to 2801 steps), see figure 4. However, the Leafcutter and the Sequential models both do much better by depleting the best source first, they get 90% and 68% in their initial surge ($\approx$ the first 100 timesteps). The Leafcutter model beats the Sequential

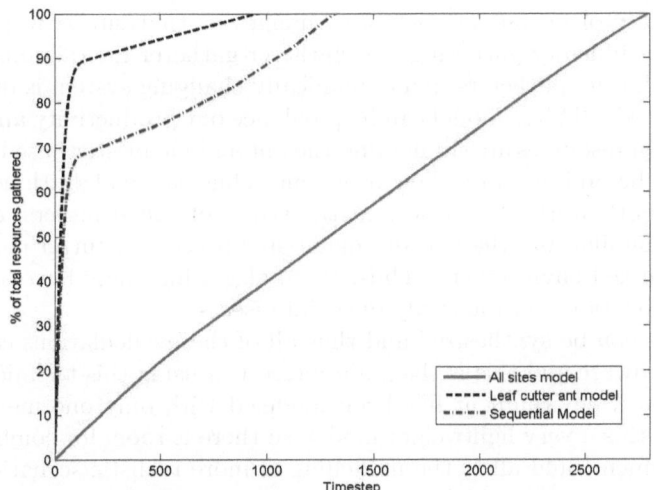

Fig. 4. How different models fare in the resource poor environment. Despite being based on ant behaviour in a rich environment, the Leafcutter ant model fares best.

model on the time taken to deplete the entire environment by 967 timesteps to 1321 and is overall more efficient. This is surprising as the Leafcutter model was based on the ant behaviour in a rich environment. Real Leafcutter ants would spread their focus amongst all the sites in a poor environment, this suggests that in such a situation ants are allocated differently to All Site allocation model.

The Leafcutter allocation model is clearly better than the Sequential model despite utilising the All Sites model over the latter time steps. This suggests that when the food sources are all of a similar magnitude (the rich environment was arguably close to this), it is better to send gatherers to all of them, weighted to the smallest, easiest depleted sources. If the sites are similar to each other this results in productivity gains whereby the slowing of productivity at one site is compensated by the increase at another. When the resource sites are vastly different, equalising the resource influx limits the best resource to the speed of gathering from the worst.

5   Conclusions

There are a few competing considerations when deciding how to spread gatherers around. More gatherers will get more resources from a single site, and the fewer resource sites there are the more gatherers can be put on each one. Thus, depleting the smaller sites first will give more gatherers to the pool to be reassigned to another resource site, increasing efficiency when the resources are similar in size. However, the extra amount of work added per gatherer decreases with the

82   E. Gale, B. de Lacy Costello, and A. Adamatzky

number of gatherers, so it makes sense to spread the gatherers around the sites. Finally, the rate of output at each site changes as that site is worked due to an increase in difficulty gathering and gatherer-gatherer interactions. Knowing how best to allocate gatherers in a dynamically changing system is difficult: the memristor based All Sites model can help balance out productivity and speed up gathering when resources are similar and the ant and memristor based Leafcutter model can be helpful in a mixed environment. Thus, these algorithms may have some use in both work allocation and the study of social insects. ering when resources are similar and the ant and memristor based Leafcutter model can be helpful in a mixed environment. Thus, these algorithms may have some use in both work allocation and the study of social insects.

Memristors can be synthesized and thus all of these calculations can be done quickly in hardware and this is the main interest in using this technique. In this paper a complex environment has been modeled with only one memristor per path, making this a very lightweight model, so there is room for combinations of memristors which could allow the modelling of more realistic scenarios. Design-wise, gathering agents could be controlled via a central memristor computer which can send out details of which path to follow (and it would make sense to have the memristors connected to reprogramable hardware so the environmental model could be changed). Alternatively, the gatherers could have the memristors physically part of them modeling the environment and then using a random number generator weighted the probabilities by the memristor voltages as to which path to go on.

The similarity between memristors and biological components has been well observed. There may also be similarities between memristors and higher levels of biological organisation. It is known that many ants can produce more complex behaviour than study of a single ant might lead us to believe and study of this phenomena has been extended to our own species [17]. Ant colonies have short-term collective memories, their complex patterns of pheromones act as stored information, so it is possible that memristors could be used to model this institutional memory. A new model for the ant's behaviour that focuses on institutional memory might shed light on the more complex institutional memory which differentiates human societies and separates us from hunter-gatherer societies.

Further Work. This is a part of ongoing work and we have been synthesizing, characterising and testing memristors, as we intend to use them to test build this model in hardware out and demonstrate that it matches the software model. Modeling pathways by a combination of one or more memristors and resistors would probably allow for a deeper investigation of effects that have been combined in this model, such as amount of resources and length of path, we intend to investigate this, and possibly back up these more detailed models with further laboratory experiments.

Acknowledgments. E.G. would like to acknowledge support from EPSRC grant EP/H014381/1.

References

1. Chua, L.O.: Memristor — The Missing Circuit Element. IEEE Trans. Circuit Theory 18, 507–519 (1971)
2. Strukov, D.B., Snider, G.S., Stewart, D.R., Williams, R.S.: The Missing Memristor Found. Nature 453, 80–83
3. Bernabé, L.-B., Teresa, S.-G.: Memristance can explain Spike-Time-Dependent-Plasticity in Neural Synapses. Available from Nature Precedings (2009), http://hdl.handle.net/10101/npre.2009.3010.1
4. Mullins, J.: Memristor minds: The future of artificial intelligence. New Scientist, 2715
5. Chua, L.O., Kang, S.M.: Memristive Devices and Systems. PIEEE 64, 209–223 (1976)
6. Howard, D., Gale, E., Bull, L., De Lacy Costello, B., Adamatzky, A.: Evolving Spiking Networks with Variable Memristors. In: GECCO 2011: Proceedings of the Genetic and Evolutionary Computation Conference. ACM Press (2011)
7. Howard, D., Gale, E., Bull, L., De Lacy Costello, B., Adamatzky, A.: Towards Evolving Spiking Networks with Memristive Synapses. In: Proceedings of the IEEE Symposium on Artificial Life. IEEE (2011)
8. Zamarreno-Ramos, C., Camunas-Mesa, L.A., Perez-Carrasco, J.A., Masquelier, T., Serrano-Gotarredona, T., Linares-Barranco, B.: On spike-timing-dependent-plasticity memristive devices, and building a self-learning visual cortex. Frontiers in Neuroscience 5, 26.1–26.22 (2011)
9. Pershin, Y., Di Ventra, M.: Solving mazes with memristors: a massively-parallel approach. Available from Nature Precedings (2011), http://dx.doi.org/10.1038/npre.2011.5748.1
10. Tarassenko, L., Blake, A.: Analogue Computation of Collision-free paths. In: Proc. IEEE: International Conference on Robotica and Automation, pp. 540–545 (1991)
11. Marshall, G.F., Tarassenko, L.: Robot path planning using resistive grids
12. Lubbock, J.: Observations on Ants, Bees, and Wasps; with a Description of a new Species of Honey-Ant. Part VII. Ants. Journal of the Linnean Society of London, Zoology 15, 167–187 (1880)
13. Wilson, E.O.: Behavior of Daceton armigerum (Latreille), with a classification of self-grooming movements in ants. Bulletin of the Museum of Comparative Zoology of Harvard College 127, 403–421 (1962)
14. Jackson, D.E., Martin, S.J., Ratnieks, F.L.W., Holcombe, M.: Spatial and temporal variation in pheromone composition of ant foraging trails. Behavioral Ecology 18, 444–450 (2007)
15. Jackson, D.E., Martin, S.J., Holcombe, M., Ratnieks, F.L.W.: Longevity and detection of persistent foraging trails in Pharaoh's ants, Monomorium pharaonis (L.). Animal Behaviour 71, 351–359 (2006)
16. Robinson, E.J.H., Jackson, D.E., Holcombe, M., Ratnieks, F.L.W.: 'No entry' signal in ant foraging. Nature 438, 442 (2005)
17. Sumpter, D.J.T.: The principles of collective animal behaviour. Phil. Trans. R. Soc. B 361, 5–22 (2006)
18. Jackson, D.E., Holcombe, M., Ratnieks, F.L.W.: Trail geometry gives polarity to ant foraging networks. Nature 432, 907–909 (2004)
19. Dorigo, M.M., Colorni, A.: Ant System: Optimization by a colony of cooperating agents. IEEE Trans. Sys. Man and Cybernetics B 26, 29–41 (1996)

20. Dorigo, M., Di Caro, G.: The Ant Colony Metaheuristic. In: New Ideas in Optimization, pp. 11–32 (1999)
21. Dorigo, M., Di Caro, G., Gambardella, M.: Ant algorithms for discrete optimization. Artificial Life 5, 137–172 (1999)
22. Stutzle, T., Hoos, H.H.: Max-min ant system. Future Generation Computer Systems 16, 889–914 (2000)
23. Bullnheimer, B., Hartl, R.F., Strauss, C.: A new rank-based version of hte Ant system: A computational study. Central European Journal for Operations Research and Economics 7, 25–38 (1996)
24. Mullen, R.J., Monekosson, D., Barman, S., Remagnino, P.: Expert Systems and Applications 36, 9608–9617 (2009)
25. Dorigo, M., Gambardella, L.M.: Ant colonies for the travelling salesman problem. BioSystems 43, 73–81 (1997)
26. Traniello, J.F.A.: Foraging Strategies of Ants. Ann. Rev. Entomol. 34, 191–210 (1989)
27. Taylor, F.: Foraging behavior in ants: experiments with two species of myrmecine ants. Behav. Ecol. Socio-biol. 2, 147–168 (1978)
28. Rockwood, L., Hubell, S.P.: Host-plant selection, diet diversity and optimal foraging in a tropical leafcutting ant. Oecologia 74, 55–61 (1987)
29. Gordon, D.M.: Ants at Work: how an insect society is organized. Free Press, Simon and Schuster (1999)

Genetic Channel Capacity Revisited

Félix Balado

School of Computer Science and Informatics
University College Dublin
Belfield Campus, Dublin 4, Ireland
felix@ucd.ie

Abstract. We revisit previous analyses on the computation of the maximum mutual information between a genetic sequence and its mutated versions down the generations, taking into account the protein translation mechanism of the genetic machinery. This amounts to the application of Shannon's capacity to the study of the transmission of genetic information. Studies on this subject were started by Yockey and then followed by a number of researchers. Here we refine prior analyses employing the Kimura model of base substitution mutations, which is more realistic than the Jukes-Cantor model used by all previous research on this topic. Furthermore we undertake exact computations where prior works just used approximations, and we propose two practical applications of genetic capacity.

1 Introduction

The origin of the application of Shannon's information theoretical concepts to molecular genetics harks back to the work of Quastler [1]. However it was Yockey who led research into this matter for many decades (a compendium of his work can be found in [2]). Yockey has compellingly argued that since information is central to the workings of molecular biology, many aspects of this discipline cannot be fully understood without the help of information theory. More recently Battail [3] has also made a good point of the relevance of information theory for unveiling the workings of evolution.

Information theory relies on probabilistic descriptions of information. For this reason, at the heart of information theoretical explanations of life lie mutations: inescapable random changes undergone by the genomes of organisms, which either go extinct or are accumulated down subsequent generations. As pointed out by Yockey, and by other researchers such as Guiaşu [4], Battail [5], May [6] and Gong et al [7], Shannon's channel capacity [8] is the fundamental limit on how well genetic information can be transmitted in front of mutations. Indeed Shannon's limit applies to any coding strategy —even if produced by evolution by natural selection. Unfortunately the bioinformatics field has paid relatively little attention to date to this potentially important research topic.

Among the aforementioned researchers, only Yockey, Guiaşu, and Gong et al analysed capacity when the protein translation mechanism of the genetic machinery is taken into account, that is, the capacity of the coding regions of genomes.

E. Hart et al. (Eds.): BIONETICS 2011, LNICST 103, pp. 85–98, 2012.
© Institute for Computer Sciences, Social Informatics and Telecommunications Engineering 2012

A number of approximations were made in previous works that leave room for further development. In this work we will put prior research in a common framework, and we will refine it by using a more realistic mutation model and by undertaking exact analyses where previous works used approximations. We will also propose two applications of channel capacity in bioinformatics research.

1.1   Notation and Basic Concepts

Calligraphic letters $(\mathcal{X})$ denote sets; $|\mathcal{X}|$ is the cardinality of $\mathcal{X}$. Boldface Roman letters $(\mathbf{x})$ denote row vectors, $\mathbf{x} = [x_1, \cdots, x_N]$. $\mathbf{1}$ is an all-ones vector. Greek capital letters $(\Pi, \boldsymbol{\Pi})$ denote matrices, and $(\Pi)_{i,j}$ is the entry of Π indexed by (i, j). $(\cdot)^T$ denotes vector or matrix transposition. A Roman letter that appears in uppercase (X) and in lowercase (x) denotes a random variable and a realisation of it, respectively. $p(X = x)$, or just $p(x)$ when unambiguous from the context, is the probability mass function (pmf) or distribution of X. For simplicity, X or the vector $\mathbf{p}_X = [p(X = x)]$ can also denote its distribution depending on the context. $p(X = x | Y = y) = p(x|y)$ denotes a conditional probability. $H(X) = -\sum_{x \in \mathcal{X}} p(x) \log p(x)$ is the entropy of a random variable X with support in $\mathcal{X}$, and $H(X|Y) = -\sum_{y \in \mathcal{Y}} p(y) \sum_{x \in \mathcal{X}} p(x|y) \log p(x|y)$ is the entropy of X conditioned to Y. $I(X;Y) = H(X) - H(X|Y)$ is the mutual information between X and Y. All logarithms are base 2 throughout the paper. The Hamming distance between $\mathbf{x}$ and $\mathbf{y}$ is denoted by $d_H(\mathbf{x}, \mathbf{y})$.

We will summarise next some basic facts about the genetic machinery that we will need in our analysis. The DNA alphabet $\mathcal{X} \triangleq \{A, C, T, G\}$ is formed by the symbols corresponding to its four bases: adenine, cytosine, thymine, and guanine. The nucleotide bases in $\mathcal{X}$ belong to two different chemical categories, namely, purines $\mathcal{R} \triangleq \{A, G\}$ or pyrimidines $\mathcal{Y} \triangleq \{C, T\}$. A codon —the minimum biologically meaningful "codeword"— is formed by a triplet of consecutive bases in a genetic sequence. A gene is formed by a sequence of codons[1] that can be translated into a sequence of amino acids, which are assembled in the same order imposed by the codons to form a protein. Using their standard short names, the set of amino acids can be written as

$$\mathcal{X}' \triangleq \{\text{Ala, Arg, Asn, Asp, Cys, Gln, Glu, Gly, His, Ile, Leu, Lys, Met, Phe,}$$
$$\text{Pro, Ser, Thr, Trp, Tyr, Val, } Stp\}, \tag{1}$$

and therefore $|\mathcal{X}'| = 21$. Every single codon $\mathbf{y} = [y_1, y_2, y_3] \in \mathcal{X}^3$ can be mapped to a unique amino acid or start/stop translation symbol, that is,

$$\xi(\mathbf{y}) = y' \in \mathcal{X}', \tag{2}$$

where the mapping $\xi(\cdot) : \mathcal{X}^3 \to \mathcal{X}'$ is established by the nearly-universal *genetic code* (easily found elsewhere, see for instance [9]), which partitions $\mathcal{X}^3$ into $|\mathcal{X}'|$

---

[1] Our analysis will be independent of the fact that a gene may not always be an unbroken sequence of codons, but rather the intertwining of noncoding sections (introns) with coding sections (exons) in the genomes of eukaryotic cells.

disjoint subsets of codons. The subset of synonymous codons associated to amino acid $y' \in \mathcal{X}'$ is $\mathcal{S}_{y'} \triangleq \{\mathbf{y} \in \mathcal{X}^3 | \xi(\mathbf{y}) = y'\}$. For instance, with our notation $\mathcal{S}_{\text{Ala}} = \{[G,C,A], [G,C,C], [G,C,T], [G,C,G]\}$ and $\xi([G,C,A]) = $ Ala. Note that the ensemble of stop codons is collected under the label *Stp* in (1). *Stp* is loosely classed as an "amino acid" for notational convenience, although it just indicates the end of gene translation and thus does not actually stand for any amino acid. Also Met (and two codons associated to Leu in eukaryotic cells) double as gene translation start symbols. We call the number of codons that map to amino acid y' the *multiplicity* of y'; this is just the cardinality of $\mathcal{S}_{y'}$, that is, $|\mathcal{S}_{y'}|$. Multiplicities are uneven over the set of amino acids, as $|\mathcal{S}_{y'}| \in \{1, 2, 3, 4, 6\}$. Due to the uniqueness of the codon-to-amino acid mapping, $\mathcal{S}_{y'} \cap \mathcal{S}_{w'} = \emptyset$ for $y' \neq w' \in \mathcal{X}'$, and $\sum_{y' \in \mathcal{X}'} |\mathcal{S}_{y'}| = |\mathcal{X}|^3 = 64$ since $\cup_{y' \in \mathcal{X}'} \mathcal{S}_{y'} = \mathcal{X}^3$.

To sum up, the main notational conventions in what follows are that regular Roman letters (i.e. y, Y), bold Roman letters (i.e. $\mathbf{y}$, $\mathbf{Y}$), and primed Roman letters (i.e. y', Y') are associated to bases, codons, and amino acids, respectively. A Greek capital letter which is both bold and primed, for instance $\boldsymbol{\Pi}'$, is a matrix associated to both codons and amino acids.

1.2 Genetic Capacity

As Shannon [8] demonstrated, the maximum mutual information between the input and the output of a channel sets the upper limit on the rate of errorless information transmission for any coding strategy over that channel —crucially, whether the coding strategy is man-made or not. Shannon called this amount *channel capacity*. In molecular genetics contexts we may assimilate mutations — and even natural selection, as we will discuss in Section 2.3— to a probabilistic channel, and we may thus talk of *genetic (channel) capacity*.

The evolution of protein synthesis marks a prominent watershed in the history of life. The mechanism to translate genes into proteins has to be taken into account from that point onwards for the definition of genetic channel capacity to be biologically meaningful. For this reason the relevant changes to be considered in the study of genetic capacity are the changes between a codon $\mathbf{Y} \in \mathcal{X}^3$ in a given DNA sequence and the amino acid $Z'_{(m)} \in \mathcal{X}'$ it translates to after m generations of the corresponding organism, that is, $Z'_{(m)} = \xi(\mathbf{Z}_{(m)})$, where $\xi(\cdot)$ is given by (2) and $\mathbf{Z}_{(m)}$ is the mutated version of $\mathbf{Y}$ after m generations of the organism (see Section 2.2). Alternatively we may also study the genetic capacity between $Y' = \xi(\mathbf{Y})$ and $Z'_{(m)}$. For this reason it is possible to contemplate two different but closely related genetic channels:

1. The *genome-proteome channel*, which is a mutation channel featuring codons at its input and amino acids at its output, and whose capacity is formulated as

$$C' \triangleq \max_{\mathbf{Y}} I(Z'_{(m)}; \mathbf{Y}) \text{ bits/amino acid.} \quad (3)$$

An alternative name for this channel given by Guiaşu is *DNA-to-protein communication channel* [4]. *DNA-mRNA-proteome communication system* was the name used by Yockey [10], who was the first to study (3).

2. The *proteome-proteome channel*, which features amino acids both at its input and at its output and whose capacity is therefore expressed as

$$C'' \triangleq \max_{Y'} I(Z'_{(m)}; Y') \text{ bits/amino acid.} \tag{4}$$

This channel was called *protein communication channel* by Gong et al [7], who were the first to study (4).

1.3   Applications of Genetic Capacity

Previous works on the computation of (3) and (4) did not discuss practical applications of Shannon's capacity in bioinformatics research. Although this is not the main purpose of this work either, we will suggest next two such applications. In the discussion that follows we will use the fact that C'' (and C') must be monotonically nonincreasing on m —as we will prove in Section 3.

1. The information content of a gene is given by its entropy $H(Y')$ (bits/amino acid). This amount can be estimated for a given gene using its empirical distribution of amino acids. According to Shannon's definition of capacity, if $\overline{m}$ is the maximum integer such that $C''|_{m=\overline{m}} = H(Y')$ then the information content of that gene will only be able to survive *unchanged* for at most $\overline{m}$ generations. A looser bound can be obtained with C'. The main consequence is that the genetic capacity establishes a quantifiable "window" of stability for genes. Qualitatively, we can infer that there should exist an evolutionary pressure towards the decrease of $H(Y')$ in successful genes, since this should increase their stability. This may be one reason why genes with uniform Y' do not exist in nature, since uniformity maximises entropy. If the mutation rate were constant and new genes were not created through duplication, a relatively lower value of $H(Y')$ would indicate a relatively older gene.
2. A number of authors have proposed to tackle phylogeny reconstruction by means of empirical estimates of the mutual information between pairs of genes (see for instance [11]). One may establish upper bounds on the branch lengths of the phylogenetic trees thus obtained by means of the genetic capacity, since it must hold that at most $\overline{m}$ generations have elapsed between two genes characterised by Y' and Z' if $\overline{m}$ is the maximum integer such that $C''|_{m=\overline{m}} = I(Z'; Y')$. Finally we must mention that the chain rule for information [12] appears not to have been yet exploited in order to build phylogenetic trees using more than two genes at a time.

2   Mutation Model

Probabilistic mutation models are required in order to evaluate (3) and (4), that is, we need to establish the equivalent of a channel model in digital communications for both cases.

2.1   Low-Level Mutation Model

Our probabilistic models will be built upon a low-level mutation channel describing a channel with bases both at its input and at its output.

Genome-Genome Channel. We denote the low-level model with this name in order to distinguish it from the high-level models that we will establish in Section 2.2. A simple but realistic model of base substitution mutations (or point mutations) is the Kimura model of molecular evolution [13]. In this model the probability that a base $y \in \mathcal{X}$ mutates to another base $z \in \mathcal{X}$ in the next generation depends on whether the mutation is a transition (that is, an intraclass mutation in which either $z, y \in \mathcal{Y}$ or $z, y \in \mathcal{R}$) or a transversion (that is, an interclass mutation in which $y \in \mathcal{Y}$ and $z \in \mathcal{R}$, or vice versa). The corresponding $|\mathcal{X}| \times |\mathcal{X}|$ base-base transition probability matrix $\Pi \triangleq [p(Z = z | Y = y)]$, with $z, y \in \mathcal{X}$, presents the following structure:

$$
\Pi = \begin{array}{c c c c c}
 & \text{A} & \text{C} & \text{T} & \text{G} \\
\begin{bmatrix}
1 - q & \frac{\gamma}{3}q & \frac{\gamma}{3}q & (1 - \frac{2\gamma}{3})q \\
\frac{\gamma}{3}q & 1 - q & (1 - \frac{2\gamma}{3})q & \frac{\gamma}{3}q \\
\frac{\gamma}{3}q & (1 - \frac{2\gamma}{3})q & 1 - q & \frac{\gamma}{3}q \\
(1 - \frac{2\gamma}{3})q & \frac{\gamma}{3}q & \frac{\gamma}{3}q & 1 - q
\end{bmatrix} & \begin{array}{c} \text{A} \\ \text{C} \\ \text{T} \\ \text{G} \end{array}
\end{array}
\tag{5}
$$

From this definition, the probability (or rate) of base substitution mutation per generation is just

$$
p(Z \neq y | Y = y) = \sum_{z \neq y} p(Z = z | Y = y) = q,
\tag{6}
$$

for any $y \in \mathcal{X}$. It must hold that $0 \leq \gamma \leq 3/2$ so that all entries of Π are probabilities; in practice $0 < \gamma < 3/2$, since $\gamma = 0$ and $\gamma = 3/2$ would forbid transversion and transition mutations, respectively. The mutation model (5) can incorporate any given transition/transversion ratio ε by setting $\gamma = 3/(2(\varepsilon + 1))$. Estimates of ε ranging between 0.89 and 18.67 for RNA and DNA of different organisms are given in [14], which correspond to γ between 0.07 and 0.79. This reflects the fact that transitions are much more likely than transversions, that is, $\varepsilon > 1/2$ virtually always in every organism, and therefore $\gamma < 1$.

The Jukes-Cantor model of molecular evolution, in which all off-diagonal elements of the transition matrix are $q/3$, is obtained as the particular case $\Pi|_{\gamma=1}$ of the Kimura model (5). Notice that this is a *mutation-symmetric model*, since its treatment of mutations between any two different bases is always the same, regardless of the category they belong to (that is, either $\mathcal{R}$ or $\mathcal{Y}$). This less realistic model is the one used in previous computations of (3) and (4) [2,7]. We will see in Section 3 that the use of the full Kimura model can lead to substantial changes in genetic capacity.

As in prior analyses we will assume that substitution mutations are mutually independent, so that the genetic channel can be taken to be memoryless. When considering m generations of an organism (that is, m cascaded mutation stages) we have a Markov chain $Y \to Z_{(1)} \to Z_{(2)} \to \cdots \to Z_{(m)}$, and model (5) leads to the overall transition probability matrix $\Pi^m = [p(Z_{(m)} = z|Y = y)]$. This matrix exponentiation can be easily calculated, either directly or by means of diagonalisation techniques.

Note that the model that we are using assumes for simplicity that the base mutation rate (6) stays constant both over the generations and over genetic loci. Although this is the standard practice in this type of analysis it stays a contentious matter. Also, in coding sequences mutation rates vary across the three positions in a codon [9], which will be further discussed in Section 2.3. Our constancy assumption may be cast as a worst-case scenario through a judicious choice of the parameter q. Finally, although substitutions are just one type of mutations, the results that we will obtain using the proposed model constitute an absolute upper bound to genetic capacity when other types of mutations — such as insertions and deletions— are factored in. We must also note that exact capacity analyses of channels with insertions and deletions are still unavailable, even in digital communications settings.

2.2   High-Level Mutation Models

Our high-level models will depend on the extension of the genome-genome channel to describe a channel with codons both at its input and at its output.

Extended Genome-Genome Channel. This channel is modelled by the $|\mathcal{X}|^3 \times |\mathcal{X}|^3$ codon-codon transition probability matrix $\boldsymbol{\Pi} \triangleq [p(\mathbf{Z} = \mathbf{z}|\mathbf{Y} = \mathbf{y})]$ associated to (5), which is just

$$\boldsymbol{\Pi} = \Pi \otimes \Pi \otimes \Pi, \tag{7}$$

where $\otimes$ is the Kronecker product[2]. This is because $p(\mathbf{Z} = \mathbf{z}|\mathbf{Y} = \mathbf{y}) = \prod_{i=1}^{3} p(Z_i = z_i|Y_i = y_i)$ according to our independent mutations assumption. Model (7) can be used to characterise any step in the Markov chain $\mathbf{Y} \to \mathbf{Z}_{(1)} \to \mathbf{Z}_{(2)} \to \cdots \to \mathbf{Z}_{(m)}$. Therefore when m such generations are considered the overall transition matrix is $\boldsymbol{\Pi}^m = [p(\mathbf{Z}_{(m)} = \mathbf{z}|\mathbf{Y} = \mathbf{y})]$, which is just $\boldsymbol{\Pi}^m = (\Pi \otimes \Pi \otimes \Pi)^m = \Pi^m \otimes \Pi^m \otimes \Pi^m$ [15].

Genome-Proteome Channel. This channel is modelled by a $|\mathcal{X}|^3 \times |\mathcal{X}'|$ codon-amino acid transition probability matrix $\boldsymbol{\Pi}' \triangleq [p(Z' = z'|\mathbf{Y} = \mathbf{y})]$. If we define a $|\mathcal{X}|^3 \times |\mathcal{X}'|$ *projection matrix* $\Lambda \triangleq [p(Z' = z'|\mathbf{Z} = \mathbf{z})]$ with entries[3]

$$(\Lambda)_{\mathbf{z},z'} = \begin{cases} 1 & \text{if } \mathbf{z} \in \mathcal{S}_{z'} \\ 0 & \text{otherwise} \end{cases}, \tag{8}$$

---

[2] $\Pi \otimes \Pi = \begin{bmatrix} (\Pi)_{1,1}\Pi & \cdots & (\Pi)_{1,4}\Pi \\ \vdots & \ddots & \vdots \\ (\Pi)_{4,1}\Pi & \cdots & (\Pi)_{4,4}\Pi \end{bmatrix}$.

[3] For notational simplicity, a matrix entry is indexed here using a codon-amino acid pair, rather than a pair of integer indices. The amino acid ordering depends on the ordering of $\mathcal{X}'$, whereas the codon ordering depends on the ordering of $\mathcal{X}$. Any such ordering is arbitrary and leads to completely equivalent channels, that is, featuring the same capacity.

then we can simply write the required matrix as

$$\boldsymbol{\Pi}' = \boldsymbol{\Pi}\,\Lambda, \tag{9}$$

where $\boldsymbol{\Pi}$ is given by (7). Of course $\Lambda 1^T = 1^T$, so that $\boldsymbol{\Pi}'$ is properly defined. This transition probability matrix becomes $\boldsymbol{\Pi}'_{(m)} = \boldsymbol{\Pi}^m\,\Lambda$ after m generations.

Proteome-Proteome Channel. This channel is modelled by a $|\mathcal{X}'| \times |\mathcal{X}'|$ amino acid-amino acid transition probability matrix $\boldsymbol{\Pi}'' \triangleq [p(Z' = z'|Y' = y')]$. If we define a $|\mathcal{X}'| \times |\mathcal{X}'|$ *multiplicity matrix* $\Omega \triangleq \Lambda^T \Lambda$, that is, a diagonal matrix for which $(\Omega)_{z',z'} = |\mathcal{S}_{z'}|$ for all $z' \in \mathcal{X}'$, then $\boldsymbol{\Pi}''$ can be written in terms of the codon-amino acid matrix (9) or in terms of the codon-codon matrix (7) as

$$\boldsymbol{\Pi}'' = \Omega^{-1}\Lambda^T \boldsymbol{\Pi}'$$
$$= \Omega^{-1}\Lambda^T \boldsymbol{\Pi}\Lambda. \tag{10}$$

Since $\Lambda^T 1^T = \Omega 1^T$ we can see that $\boldsymbol{\Pi}''$ is properly defined. When m generations are considered the transition probability matrix becomes[4] $\boldsymbol{\Pi}''_{(m)} = \Omega^{-1}\Lambda^T \boldsymbol{\Pi}^m\,\Lambda$. See that $\Omega^{-1}\Lambda^T$ is just the $|\mathcal{X}'| \times |\mathcal{X}|^3$ *codon usage matrix* $\Upsilon \triangleq [p(\mathbf{Y} = \mathbf{y}|Y' = y')]$ assuming uniformity. A further increase of C'' could be achieved by optimising Υ; however we will see in Section 3 that this cannot make a great difference since C'' computed using (10) is closely upper bounded by C'.

Finally, while preparing the final version of this article it came to our attention that a similar method for obtaining $\boldsymbol{\Pi}''$ from an empirical estimate of $\boldsymbol{\Pi}$ was previously given in [16], although without the explicit matrix formulation above. In our notation, the approach in [16] uses empirical codon usage to define Υ.

2.3   Natural Selection and Probabilistic Molecular Evolution Models

Coding regions of genomes define proteins and are thus subject to much stronger selection pressures than noncoding regions, whose change can be more obviously considered to be mainly driven by random mutations. Therefore it might appear that a purely probabilistic approach to genetic information transmission would only make sense for noncoding regions, using a low-level model such as (5), whereas it would seem that the effects of natural selection are ignored in the high-level models (9) and (10).

However we must realise that natural selection may also be accounted for in a probabilistic way. This is actually the premise of the so-called *neutral theory of molecular evolution* proposed by Kimura [17]. In order to do so with our high-level models we just need to choose a base mutation rate q that combines the effect of both mutations and natural selection in coding regions. The resulting $\boldsymbol{\Pi}''$ matrices, in particular, will be akin to the so-called point accepted mutation matrices (PAM) first estimated by Dayhoff et al [18]. A PAM matrix is basically an empirical estimate of $\boldsymbol{\Pi}''$ obtained from real data —unlike our model, without

---

[4] If $\boldsymbol{\Pi}''$ were available but not $\boldsymbol{\Pi}$ one could only use $(\boldsymbol{\Pi}'')^m$ to model this case.

imposing any structure constraint. These matrices are used in the reconstruction of phylogenies and, hence, have to statistically account for evolutive pressures.

We must also remark that, if necessary, the effect of natural selection on coding regions can be reflected even more accurately in our high-level models (9) and (10). The way to do this is to take into account that natural selection implies that the base mutation rate q_w corresponding to the third base in a codon (*wobble* position) is much higher that the base mutation rate q corresponding to any of the bases in the initial duplet. This is because the genetic code implies that mutations affecting the initial duplet of a codon are more prone to induce proteomic change, and hence less likely to survive in subsequent generations. Therefore, if the relevant mutation rates are available, the codon-codon transition probability matrix (7) can be more accurately obtained as $\boldsymbol{\Pi} = \Pi|_q \otimes \Pi|_q \otimes \Pi|_{q_w}$. We will not pursue this idea further in this paper, because we will see in Section 3 that our basic model is sufficient already in order to approximate the results of a PAM model for the purpose of computing genetic capacity.

3   Genetic Capacity Computation

Equipped with the channel models in Section 2.2 we are now ready to obtain the capacities (3) and (4) of the channels that describe the information flow from the genome to the proteome and from the proteome to the proteome, respectively. We will also retrace in this section previous studies on the matter before computing C' and C'' by relying on our Kimura-based models (9) and (10)

In general, the channel matrix $\boldsymbol{\Pi}'_{(m)}$ required to compute (3) does not correspond to a symmetric[5] (or weakly symmetric[6]) channel, which would evince uniform $\mathbf{Y}$ as the maximising input distribution. Similarly the channel defined by $\Pi''_{(m)}$ is not symmetric either, and thus uniform Y' does not necessarily lead to capacity in (4). In spite of this, the required optimisations can be easily undertaken numerically by means of the Blahut-Arimoto algorithm [19]. If $\mathbf{p_Y} = [p(\mathbf{Y} = \mathbf{y})]$ is the pmf that yields C' in (3) observe that we may not always be able to find a pmf $\mathbf{p}_Y = [p(Y = y)]$ such that $\mathbf{p}_Y \otimes \mathbf{p}_Y \otimes \mathbf{p}_Y = \mathbf{p_Y}$. This means that (3) and (4) are upper bounds with respect to maximising the corresponding mutual informations on Y (that is, the input distribution of bases), which is not so straightforward but not our goal here either.

Before any actual computation, one can already see that both C' and C'' have to be monotonically nonincreasing on m. This is because we can establish the Markov chain $Y' \to \mathbf{Y} \to Z'_{(1)} \to \cdots \to Z'_{(m)}$ which implies both that $I(Z'_{(1)}; \mathbf{Y}) \geq \cdots \geq I(Z'_{(m)}; \mathbf{Y})$ and that $I(Z'_{(1)}; Y') \geq \cdots \geq I(Z'_{(m)}; Y')$ by repeatedly applying the data-processing inequality [12]. We can also see that $C' \geq C''$ for any given m, as the same Markov chain similarly implies that $I(Z'_{(m)}; \mathbf{Y}) \geq I(Z'_{(m)}; Y')$.

---

[5] All rows the transition probability matrix are permutations of the same set of probabilities, and so are its columns.

[6] All rows of the transition probability matrix are permutations of the same set of probabilities, and all its columns add up to the same number.

Genome-Proteome Capacity. As regards prior work on this subject, it is worth mentioning the contribution by Guiaşu [4] in the first place, although this is not the first work dealing with (3). Guiaşu's analysis is for $m = 1$, and uses the transition probability matrix $\boldsymbol{\Pi}'|_{q=0} = \boldsymbol{\Pi}|_{q=0} \Lambda = \Lambda$ which corresponds to a mutation-free scenario, that is, $I(Z'; \mathbf{Y}) = H(Z') = H(Y')$. The maximum in this case can be analytically obtained by using any distribution of $\mathbf{Y}$ that makes Y' uniform, that is, $\mathbf{p_Y}\Lambda = 1/|\mathcal{X}'|$, for instance $\mathbf{p_Y} = (1\Omega^{-1}\Lambda^T)/|\mathcal{X}'|$. Guiaşu observed that the corresponding maximum $C'|_{q=0} = \log|\mathcal{X}'| = 4.39$ bits/amino acid is not observed in nature, and he also speculated about (3) when $q > 0$.

This case was actually first tackled by Yockey [10] for $m = 1$ —Guiaşu's work was a later but apparently independent development. Yockey assumed the Jukes-Cantor model of substitution mutations, which he referred to as "white genetic noise", and used an approximation $\widetilde{\boldsymbol{\Pi}}$ of $\boldsymbol{\Pi}|_{\gamma=1}$ in which all entries that correspond to more than one base mutation per codon are nulled. An analytical expression of $\widetilde{\boldsymbol{\Pi}}$ can be obtained as follows. Given the sets of codons $\mathcal{I}_\mathbf{y} \triangleq \{\mathbf{z} \in \mathcal{X}^3 | d_H(\mathbf{y}, \mathbf{z}) = 1\}$ for $\mathbf{y} \in \mathcal{X}^3$, see that $|\mathcal{I}_\mathbf{y}| = (|\mathcal{X}| - 1)^3 = 9$ for any $\mathbf{y}$. Therefore Yockey's approximation is

$$(\widetilde{\boldsymbol{\Pi}})_{\mathbf{y},\mathbf{z}} = \begin{cases} \alpha & \text{if } d_H(\mathbf{y}, \mathbf{z}) = 1 \\ 1 - 9\alpha & \text{if } \mathbf{y} = \mathbf{z} \\ 0 & \text{otherwise} \end{cases}, \tag{11}$$

with $\alpha = (1 - (1 - q)^3)/9$. Yockey implicitly approximates this parameter as

$$\alpha \approx \frac{q}{3} \tag{12}$$

which is accurate for $q \ll 1$; however this approximation breaks down for $q > 1/3$ since in this range the diagonal entries of $\widetilde{\boldsymbol{\Pi}}$ must be negative for its rows to add up to one. Yockey discussed how to compute $I(Z'; \mathbf{Y})$ using the channel matrix $\widetilde{\boldsymbol{\Pi}}' = \widetilde{\boldsymbol{\Pi}} \Lambda$ (which is just Table 5.2 in [2, page 50]) and a given distribution of $\mathbf{Y}$, although he did not attempt maximisation, and he also produced the following approximation [2, page 52]

$$I(Z'; \mathbf{Y}) \approx H(\mathbf{Y}) - 1.7915 \quad 9.815\alpha$$
$$+ 34.2108\alpha^2 + 6.8303\alpha \log \alpha. \tag{13}$$

In Figure 1 we plot C' for $m = 1$ and the whole range of q using Yockey's $\widetilde{\boldsymbol{\Pi}}'$ matrix and the Kimura-based $\boldsymbol{\Pi}'|_{\gamma=1}$ matrix obtained from (9) (that is, Jukes-Cantor based). All results are computed using the Blahut-Arimoto algorithm. The capacity computed with $\widetilde{\boldsymbol{\Pi}}'$ is accurate for $q \ll 1$, but it plateaus as q increases. See that with an exact analysis $C'|_{q=3/4} = 0$ when $\gamma = 1$, since in this case $\boldsymbol{\Pi} = \mathbf{1}^T\mathbf{1}/|\mathcal{X}|^3$ and then no information can be conveyed because any codon can mutate to any other with uniform probability. C' is increasing in the range $q \in [3/4, 1]$ because the uncertainty about $\mathbf{Y}$ given Z' is smaller when $q = 1$ than when $q = 3/4$: given z' when $q = 1$, if a base z is the same for one of the three

94   F. Balado

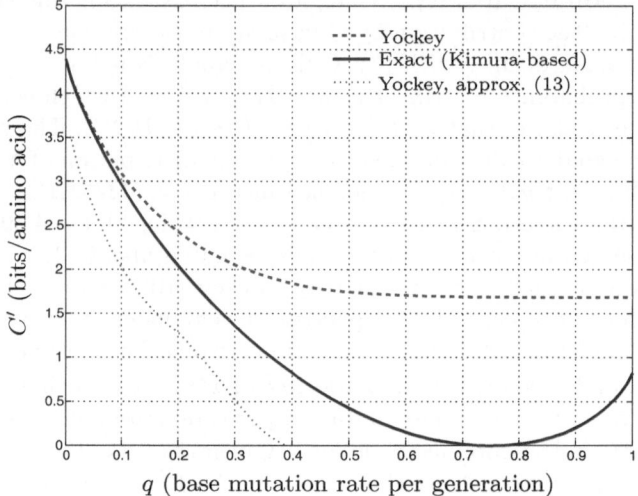

Fig. 1. Genome-proteome genetic capacity ($\gamma = 1$, $m = 1$)

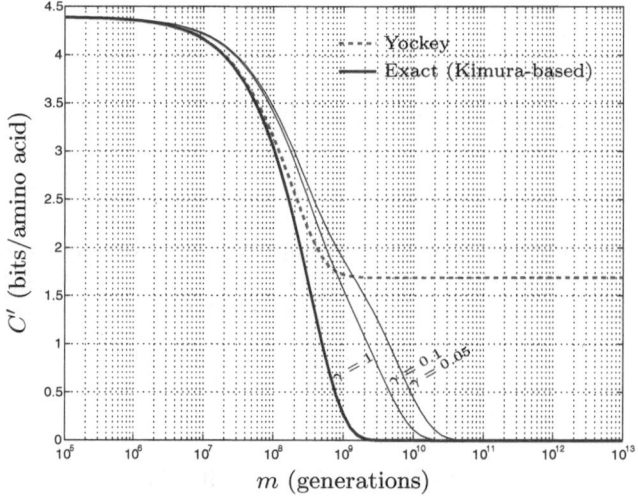

Fig. 2. Genome-proteome genetic capacity ($q = 10^{-9}$)

positions of all $\mathbf{z} \in \mathcal{S}_{z'}$ this implies that $y \neq z$ for the corresponding base in $\mathbf{Y}$ (and hence capacity is higher than when $q = 3/4$). Finally, both plots coincide at $C'|_{q=0}$ with the value given by Guiaşu. As for approximation (13), computed using the optimal distribution of $\mathbf{Y}$ obtained when applying the Blahut-Arimoto algorithm, we can see that it is not very exact: it becomes negative when $q \gtrsim 0.4$.

Yockey suggested in [2] that his model could be extended to comprise transversions and transitions, but argued that this would not make an important difference. Figure 2, which depicts C' as a function of m, shows that this assumption was not completely correct. In the realistic case $\gamma < 1$, model $\mathbf{\Pi}'_{(m)}$

obtained using (9) yields a noticeable capacity increase afforded by mutation-symmetry breaking. Furthermore Yockey's capacity, computed in Figure 2 using $\widetilde{\Pi}'_{(m)} = \widetilde{\Pi}^{m} \Lambda$, eventually deviates from the exact capacity for $\gamma = 1$ as m increases; in particular it does not tend to zero as $m \to \infty$, which limits the applicability of approximation (11) even for $q \ll 1$.

Proteome-Proteome Capacity. As in the previous section we will review and discuss previous research on this matter and compare it with our computations. Gong et al [7], who were the first to study (4), used two different mutation models in their genetic capacity analysis:

1. The first one is parallel to Yockey's. Gong et al propose a Jukes-Cantor model of base substitution mutations and put forward computational reasons to make the exact same approximation as Yockey did, that is, they disregard transitions between codons at Hamming distance greater than one. Consequently their resulting channel matrix (**P** in Figure 5 from [7], which we will denote here as $\widetilde{\Pi}''$ for the sake of keeping our notation) is essentially[7] obtained as

$$\widetilde{\Pi}'' = \Omega^{-1} \Lambda^T \widetilde{\Pi} \Lambda, \tag{14}$$

where $\widetilde{\Pi}$ is given by (11). This model is used in [7] to obtain C'' for $m = 1$ as a function of the base mutation rate q.
2. The second one is a point accepted mutation (PAM) matrix [18], that is, an unconstrained estimate of Π'' obtained from real data. This model is used in [7] to obtain C'' as a function of m.

The proteome-proteome channel is not symmetric with any of these two models, and hence Gong et al apply the Blahut-Arimoto algorithm to obtain C''. Unfortunately their **P** model suffers from the same shortcomings as Yockey's. Figure 3 shows C'' when $m = 1$ for the whole range of q, both using the approximation $\widetilde{\Pi}''$, with $\alpha = (1 - (1 - q)^3)/9$, and the exact Kimura-based model $\Pi''|_{\gamma=1}$ obtained from (10). The capacity obtained using $\widetilde{\Pi}''$ is accurate for $q \ll 1$ but plateaus around $C'' = 1.5$ bits/amino acid as q increases, which is not correct for the same reasons as Yockey's result for C' in Figure 1. We must also note that the plot corresponding to Figure 3 in [7] (Figure 6(b), where α is approximated as in (12)) shows C'' only up to $\alpha = 1/9$ (that is, $q = 1/3$) for the reason indicated when discussing the range of validity of (12).

Our next comparison is presented in Figure 4, which gives C'' as a function of m both using $\widetilde{\Pi}''_{(m)} = \Omega^{-1} \Lambda^T \widetilde{\Pi}^{m} \Lambda$ and $\Pi''_{(m)}$ computed using (10) for different values of γ. The results in this plot are once more parallel to the ones for C' (Figure 2), in particular the mutation-symmetry breaking effect is also observed. We observe again that the genetic capacity does not tend to zero when using approximation $\widetilde{\Pi}''_{(m)}$ as the number of generations m increases.

This issue of the **P** model is probably behind the use of a PAM model in [7] in order to study C''' as a function of m. With this alternative model these authors

---

[7] The channel matrices in [7] are $(|\mathcal{X}'| - 1) \times (|\mathcal{X}'| - 1)$, as they leave *Stp* out.

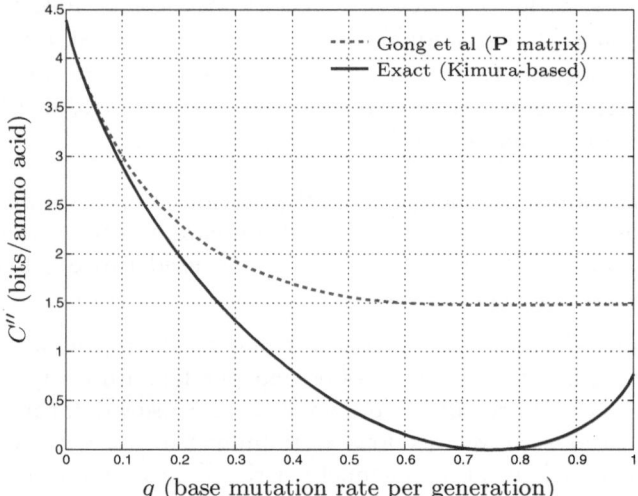

Fig. 3. Proteome-proteome genetic capacity ($\gamma = 1$, $m = 1$)

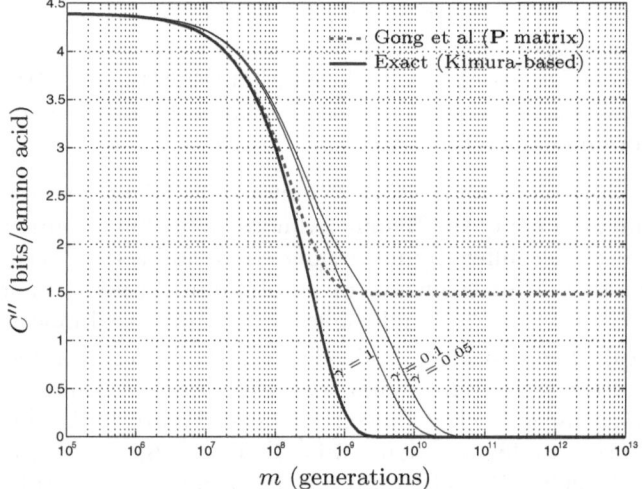

Fig. 4. Proteome-proteome genetic capacity ($q = 10^{-9}$)

able to show that $C'' \to 0$ as $m \to \infty$. However our much simpler Kimura-based model (10) is also able to closely replicate the corresponding result from [7]. To do so we just need to set $q = \frac{1}{3} \times 10^{-2}$ and $\gamma = 1$ so that Π'' represents the 1 PAM matrix given in [18]. Note that $\Pi''_{(m)} \neq (\Pi'')^m$ because $\Lambda \Omega^{-1} \Lambda^T$ is not the identity matrix, but C'' can be empirically shown to be very similar in both cases. The relevance of the comparison presented in Figure 5 is that our model only depends on two parameters, whereas a general PAM matrix requires estimating 380 parameters to model all pairwise amino acid mutabilities.

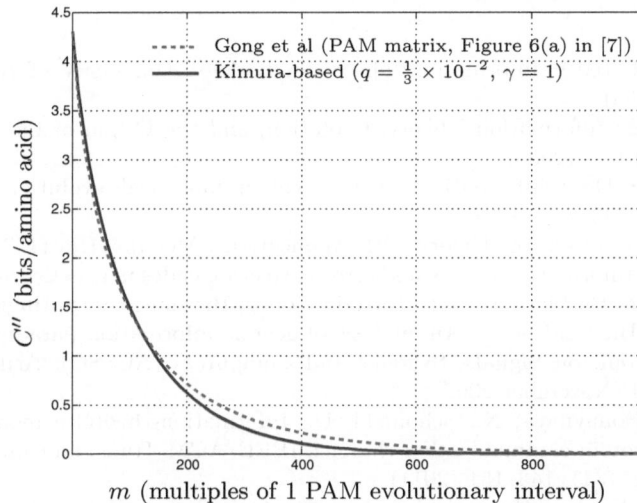

Fig. 5. Proteome-proteome genetic capacity

4 Conclusions

We have provided a reevaluation of the two genetic capacity measures that take into account the protein translation mechanism, leading to more precise and simpler analyses. We have seen that the channel approximations proposed by Yockey [2] and by Gong et al [7] yield capacity accurately for small values of q (Figures 1 and 3), but even with $q \ll 1$ they are not suitable when accumulated mutations are taken into account as $m \to \infty$ (Figures 2 and 4).

We have also shown that Shannon's capacity turns out to be greater using the more realistic Kimura model with respect to the previously used Jukes-Cantor model. The increase is roughly of one order of magnitude at $C' = 0.01$ ($C'' = 0.01$) in terms of the number of generations m for the realistic transition-transversion parameter $\gamma = 0.1$ (Figures 2 and 4). Finally, we have shown that our simple model of Π'', which depends on two parameters only, is able to closely reproduce the capacity obtained using PAM matrices (Figure 5), which require estimating hundreds of parameters from real data.

Genetic channel capacity remains largely unexplored in the bioinformatics field. We have discussed two concrete applications that could spur further research, but other suitable scenarios can easily be conceived. As a fundamental limit to elementary processes in molecular biology, genetic capacity has the potential to become a staple of practical bioinformatics investigations in the future.

Acknowledgements. This publication has emanated from research conducted with the financial support of Science Foundation Ireland under Grant Number 09/RFP/CMS2212.

98   F. Balado

References

1. Quastler, H. (ed.): Information Theory in Biology. University of Illinois Press, Urbana (1953)
2. Yockey, H.P.: Information Theory, Evolution, and the Origin of Life. Cambridge University Press (2005)
3. Battail, G.: Does information theory explain biological evolution? Europhys. Lett. 40(3), 343–348 (1997)
4. Guiaşu, S.: Information Theory with Applications. McGraw-Hill (1977)
5. Battail, G.: Information theory and error-correcting codes in genetics and biological evolution. In: Barbieri, M. (ed.) Introduction to Biosemiotics. Springer (2007)
6. May, E.E.: Bits and bases: An analysis of genetic information paradigms. In: 41st Asilomar Conf. on Signals, Systems and Computers (ACSSC), Asilomar, USA, pp. 165–169 (November 2007)
7. Gong, L., Bouaynaya, N., Schonfeld, D.: Information-theoretic model of evolution over protein communication channel. IEEE/ACM Trans. on Comp. Biol. and Bioinformat. 8(1), 143–151 (2011)
8. Shannon, C.E.: A mathematical theory of communication. Bell System Tech. J. 27, 379–423, 623–656 (1948)
9. Li, W.: Molecular Evolution. Sinauer Associates (1997)
10. Yockey, H.P.: An application of information theory to the central dogma and the sequence hypothesis. J. Theor. Biol. (46), 369–406 (1974)
11. Yu, Z., Mao, Z., Zhou, L.-Q., Anh, V.: A mutual information based sequence distance for vertebrate phylogeny using complete mitochondrial genomes. In: Procs. of the IEEE 3rd Intl. Conf. on Natural Computation, Haikou, China, pp. 253–257 (2007)
12. Cover, T.M., Thomas, J.A.: Elements of Information Theory. Wiley-Interscience (1991)
13. Kimura, M.: A simple method for estimating evolutionary rate in a finite population due to mutational production of neutral and nearly neutral base substitution through comparative studies of nucleotide sequences. J. Mol. Biol. 16, 111–120 (1980)
14. Purvis, A., Bromham, L.: Estimating the transition/transversion ratio from independent pairwise comparisons with an assumed phylogeny. J. of Mol. Evol. 44, 112–119 (1997)
15. Magnus, J.R., Neudecker, H.: Matrix Differential Calculus with Applications in Statistics and Econometrics, 3rd edn. John Wiley & Sons (1999)
16. Mackiewicz, P., Biecek, P., Mackiewicz, D., Kiraga, J., Baczkowski, K., Sobczynski, M., Cebrat, S.: Optimisation of Asymmetric Mutational Pressure and Selection Pressure Around the Universal Genetic Code. In: Bubak, M., van Albada, G.D., Dongarra, J., Sloot, P.M.A. (eds.) ICCS 2008, Part III. LNCS, vol. 5103, pp. 100–109. Springer, Heidelberg (2008)
17. Kimura, M.: The Neutral Theory of Molecular Evolution. Cambridge University Press (1983)
18. Dayhoff, M., Schwartz, R., Orcutt, B.: A model of evolutionary change in proteins. Atlas of Protein Sequence and Structure 5(3), 345–352 (1978)
19. Blahut, R.: Computation of channel capacity and rate-distortion functions. IEEE Trans. on Inf. Theory 18(4), 460–473 (1972)

Biologically Inspired Attack Detection in Superpeer-Based P2P Overlay Networks

Paul L. Snyder, Yusuf Osmanlioglu, and Giuseppe Valetto

Drexel University, Philadelphia PA 19104, USA

Abstract. We present a bio-inspired mechanism that allows a peer-to-peer overlay network to adapt its topology in response to attacks that try to disrupt the overlay by targeting high-degree nodes. Our strategy is based on the diffusion of an "alert hormone" through the overlay network, in response to node failures. A high level of hormone concentration in a node induce that node to switch protocol. That leads to a self-organized modification of the entire overlay from a superpeer, scale-free layout, to a flatter network that is much less vulnerable to targeted attacks. As the hormone is metabolized with time, nodes switch back to the original protocol and reconstruct a superpeer overlay. We demonstrate and evaluate this mechanism on top of the peer-to-peer *Myconet* overlay, which is itself self-organized and bio-inspired.

Keywords: P2P, self-organization, self-protection, bio-inspired, diffusion.

1 Introduction

Contemporary communication and computing networks are frequently very large in scale and extremely dynamic. Comprised of many nodes with widely varying capabilities, they must be able to operate effectively in the face of continually changing communities of nodes, as well as malicious activity that may attempt at any time to degrade or disrupt the network. Centralized organization and administration is thus a poor fit for these highly distributed systems.

Overlays are often used to construct a topology on top of an underlying network. By imposing a degree of order, overlays can increase the efficiency of services such as search, routing, or application-level activities. A type of overlay that is well represented in peer-to-peer (P2P) networks is that of unstructured, superpeer-based overlays. Such overlays attempt to exploit the heterogeneity of nodes in the network by selecting the nodes with more resources and better capabilities to become superpeers that take on a service roles for many other, less capable peers. Superpeers become network *hubs*, with much higher degrees than regular peers, often with the intent of building a *scale-free* topology for the overlay [1,2]. Many examples of superpeer-based P2P networks and protocols exist, including the well-known Gnutella [3], Kazaa [4], and Skype [5], as well as numerous research protocols.

In addition to an improved efficiency in leveraging the heterogeneous resources of its nodes, a superpeer overlay is also more resilient to random node failures,

E. Hart et al. (Eds.): BIONETICS 2011, LNICST 103, pp. 99–114, 2012.

compared to a non-hierarchical overlay where all nodes have similar degrees [1]. However, superpeer overlays are inherently very vulnerable to directed attacks, that is, those that specifically target superpeers. By focusing on the highest-degree superpeers, an attacker can quickly disrupt the whole network, possibly breaking it into multiple components and isolating many of the regular peers. To combat this kind of attack strategy, researchers have begun to investigate self-organizing superpeer protocols, which support the rapid reconstruction of an efficient overlay topology after a disruptive event (either malicious or accidental): examples include SG-1 [2], ERASP [6] and our own Myconet protocol, which takes inspiration from the robust structure and resilient behavior of *mycelia*, the root-like networks formed by many species of fungi [7,8].

While those self-organized, self-healing protocols can help a network recover quickly from targeted attacks, they cannot directly thwart them. One possible strategy, as suggested among others by Keyani et al. [9] and Zweig and Zimmerman [10], is to implement a provision in the network that causes nodes to switch from a superpeer, scale-free mode, to a "flat" mode, in which the network nodes maintain a uniform or narrower degree distribution. Zweig and Zimmerman [10] discuss how such provision should be reactive and fully decentralized, should rely only on information available locally to each node, and should not depend on an accurate distinction between the failure of a node due to an actual attack, as opposed to an accident. Keyani et al. [9] note that a critical property of such an attack-thwarting provision is the ability to switch back reliably to the more efficient scale-free topology, if the network is subject only to random failures due to normal levels of "node churn".

In this paper, we present a mechanism to defend against targeted attacks in superpeer overlays, which implements the protocol switch strategy discussed above. Our mechanism is biologically inspired, and is based on the release and diffusion of an "alert hormone" in the network in response to node failures. The alert hormone propagates from the neighbors of the failed node throughout the network. The local concentration of the hormone at each node is used by that node as a hint that an attack may be happening somewhere in the overlay, and, when it exceeds a threshold, induces the node to make a local decision to switch from a superpeer-oriented strategy for the overlay construction and maintenance, to a strategy in which the node seeks to remain attached to a constant, small number of peers. The diffusion of the hormone in a network under a targeted attack causes the network as a whole to adapt its topology and quickly switch to a "flat" mode, where all nodes have an almost uniform degree. In that mode, an attacker is deprived of major targets, and the network becomes harder to disrupt. As the attack subsides (or is reduced to simply killing random low-degree nodes), the hormone is metabolized by each node. When the local concentration falls below a threshold, nodes will switch their protocol back to normal operation, and start working towards re-establishing a superpeer overlay.

We have built the mechanism outlined above on top of the self-healing Myconet protocol, which is itself self-organized and biologically inspired. Myconet characteristics include rapid construction and reconstruction of superpeer overlays, and a selection policy for superpeer promotions that efficiently exploits the heterogenous capacities of nodes in the network.

In Section 2, we provide an overview of Myconet, and discuss how the alert hormone strategy for topology adaptation has been implemented on top of the basic Myconet protocol. In Section 3, we describe our evaluation and present its results, followed by a discussion of their implications in section 4, along with directions for future work. Relevant related work is examined in Section 5. Conclusions are presented in Section 6.

2 Approach

As the basis for this work, we used Myconet [7], a protocol for constructing superpeer-based peer-to-peer overlays, inspired by the growth patterns of fungal *hyphae*. We first outline the major elements of the protocol as background (a detailed description can be found in [7]; only a high level overview is presented here due to space limitations). Section 2.2 we describe the changes made to the protocol in order to implement our attack response and topology self-adapting mechanism. While our implementation leverages the characteristics of the Myconet protocol, the technique, based upon the principle of hormone diffusion, should be applicable to other superpeer-based overlays. The terms *peer* and *node* are used interchangably in this paper.

2.1 Background

In accord with the fungal metaphor, superpeers in Myconet are considered to be hyphae (or *hyphal peers*), and non-superpeer nodes as *biomass*, to be moved among the hypahe as needed. We use an abstract concept of *capacity* to model heterogenous node capability: higher-capacity nodes make better superpeers, because they are able to be connected to (and service) more of the other peers.

In Myconet, local protocol dynamics work to identify the "best" peers from a heterogeneous assortment and select the highest-capacity to become hyphal peers. Hyphal peers move through a series of protocol states and build a robust network of interconnections with other hyphal peers. They also act to heal the overlay in case of disconnection by damage, whether from incidental peer failures or malicious attacks. Peer states and transitions are shown in the diagram in Figure 1. When promoted from biomass to the first, *extending* hyphal state, superpeers are mainly focused on exploration, acting as connection points for new biomass peers entering the network. *Branching* hyphae are extending superpeers that have reached their target level of utilization (i.e., they have connected to sufficient biomass peers to reach a level of utilization appropriate to their capacity); they manage the number of extending peers in the network, while growing new hyphal interconnections. Finally, *Immobile* hyphae have reached levels of full biomass connection and a target number of hyphal interconnections, and are considered to be relatively stable, having remained in the network long enough to transition through all the previous protocol states. When hyphal peers contact

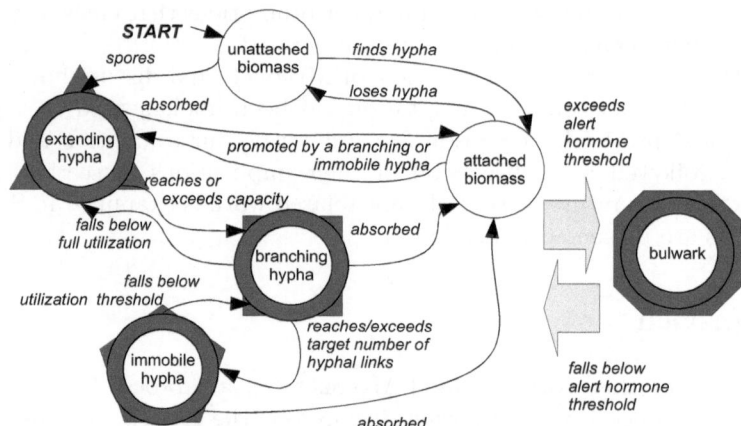

Fig. 1. Myconet protocol state transitions. States are are described in Section 2.

each other, superpeers of lower protocol state and lower capacity transfer (part of) their attached biomass peers towards higher superpeers, in order to saturate the capacity of the latter. Should a hyphal peer for any reason fall below a utilization threshold, it may demote itself to a lower protocol state, possibly reverting to become a biomass peer.

Myconet constructs and maintains a strongly interconnected, superpeer-based overlay that quickly converges to an optimal number of superpeers. It quickly self-heals, repairing damage caused by failed peers, and dynamically adjusts the network topology to changing conditions. Results from applying Myconet to a clustering and load-balancing application can be found in [8].

Myconet relies on a lower-level gossip layer as a source of random nodes and their states (as a way to find appropriate hyphal peers for connection). The current implementation uses a simplified version of the Newscast protocol [11], modified to include state information.

2.2   Protocol Modifications

Damage to the overlay caused by peer failure (whether by attack or normal churn) results in the generation of a marker (metaphorically considered to be an *alert hormone*) that spreads from neighbor to neighbor by diffusion. Once sufficient amounts of alert hormone have accumulated in a peer, it switches into a new protocol state. This state is termed *bulwark*, indicating that the peer has switched into a defensive posture and is acting to reduce the network's vulnerability to attack. Bulwark peers do not follow the regular Myconet protocol, but simply work to configure their local neighbor relationships into a flat (non-superpeer) network by maintaining a fixed, small number of connections to other peers. When a peer switches into the bulwark state, it severs connections with all its existing neighbors, causing the generation of yet more alert hormone.

Alert Hormone Generation. When a peer fails, the amount of hormone generated by the remaining peers is determined by the failed peer's degree (that is, its number of neighbors). A peer that observes the loss of a neighbor through a method other than normal Myconet protocol dynamics will generate a quantity of alert hormone according to a quadratic relationship, $0.5d^2 - 2d$, where d is the degree of the failed neighbor. This function weights heavily the failure of large degree peers (such as immobile or branching hyphal peers), whereas it is relatively insensitive to the failure of low-degree peers (such as biomass and peripheral extending hyphal peers). Since every neighbor of a failed peer will generate the alert hormone according to this formula, the failure of large-degree peers will result in the release of large amounts of hormone throughout the network.

Hormone Dynamics. Diffusion occurs through simple neighborhood equalization. If a peer p's neighbor peer n has a lower hormone concentration than p, the two nodes will balance their hormone levels. Over time, this results in the hormone diffusing through the network.

The maximum quantity of alert hormone that may be held at a node is capped, to prevent the build-up of network pockets with very large amounts of hormone. If, after diffusion, the amount of hormone is over this maximum, the excess is discharged by the peer and removed from the system.

Additionally, peers metabolize the alert hormone over time. The local concentration is periodically reduced by a fixed percentage. Thus, hormone levels will spike following a failure, but will slowly drop back down if additional failures do not continue to generate additional alert signals.

The hormone maximum quantities and decay rate are configurable parameters, discussed in more detail in Section 2.3.

Switching the Mode of the Network. A peer whose level of alert hormone is over a certain threshold *may* switch into the bulwark state, as shown in Figure 1. In order to dampen the instant effect of large spikes, the local concentration must remain over the threshold for a period of time.

When a peer makes the jump into the bulwark state, it will drop all connections with its current neighbors. This drop is abrupt, causing all the former neighbors to release alert hormone into the network. This may start a cascade of more peers transitioning to the bulwark state, depending on the existing local hormone levels. See Figure 2 for a visualization of protocol dynamics.

The choice to drop neighbors was made in order to simplify implementation (reducing the types of signals needed, as only failure detection is required) and most importantly to speed the spread of alert hormone, since cascaded bulwark nodes will connect to a different set of neighbors. A future implementation will explore the performance effects on global mode change of retaining some or all the neighbors following a node's transition to the bulwark state.

When switching, a bulwark peer caches its previous Myconet protocol state. It immediately begins to connect to new neighbors, growing new connections to other peers until it has reached a target number. If all peers enter the bulwark state, the network converges to a relatively flat topology, without superpeers. The network thus becomes resilient against targeted attacks, although, while in this mode, the network cannot effectively exploit heterogeneous node capabilities.

Once the network has shifted to a flat configuration, attacks against particular peers are practically indistinguishable from churn. Because of this, the failure of a bulwark peer does not cause additional alert hormone to be generated.

A peer will remain in the bulwark state until its local concentration of alert hormone has dropped below a threshold value, plus an additional period of time. This latency is a customizable parameter, and is designed into the protocol to prevent the network from attempting to switch between modes too quickly, which may result in the overhead of restoring the superpeer topology while waves of hormones are still traversing the network, possibly because an attacker is still trying to disrupt the overlay.

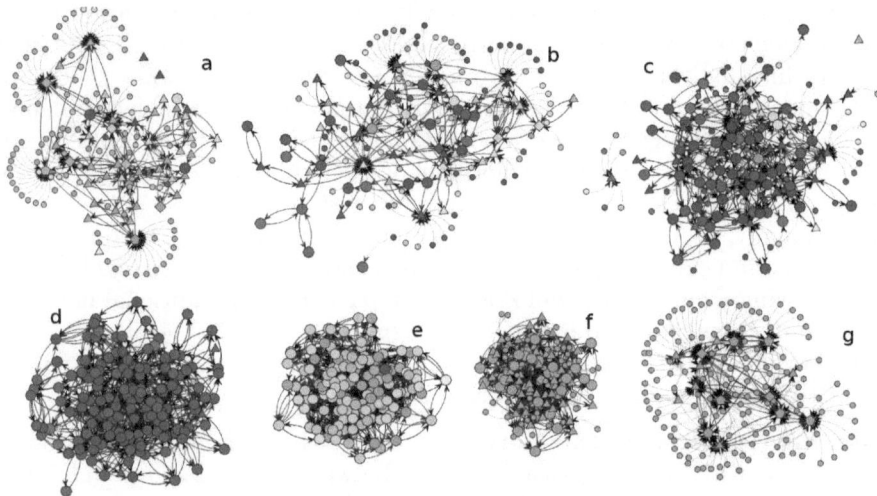

Fig. 2. Hormone-based attack detection in an example 150-node network with $c_{max} = 20$ (see Section 3.1 for a description of this parameter). (a) Two cycles into an attack, some local buildup of alert hormone is occuring. (b) Four cycles into the attack, nodes have passed the transition threshold (red) and switched into bulwark (octagons), causing more nodes to make the jump. (c) Alert hormone levels continue to rise, diffusing through all nodes. (d) Six cycles into the attack, all nodes have switched to the bulwark state. (e) Four rounds after the attack ends, hormone concentrations in most nodes have decayed below the reversion threshold; nodes delay their return by five cycles to reduce thrashing. (f) Nine cycles after the attack ends, alert hormone levels have decayed and the nodes revert to normal operation. (g) Within six more cycles, the nodes have reconstructed the superpeer overlay.

Modifications to Other Protocol States. Besides the addition of the bulwark state, all the states of the Myconet protocol were modified, in order to accommodate the new hormone-based topology self-adaptation described above.

When a peer drops out of bulwark state, it first attempts to return to the My-conet state it had cached when it switched. This provides a "hint" as the network rebuilds, speeding the identification and promotion of high-quality nodes.

Bulwark peers are not considered to be hyphal peers for purposes of counting hyphal neighbor targets (for branching and immobile peers), though they can serve as connection points for new peers entering the network (as extending peers do). They are counted as biomass when determining a peer's utilization.

Thus, any node executing the Myconet protocol may have additional neighbors in the bulwark state, above the number of neighbors that it would normally attempt to maintain; this is particularly significant for biomass nodes, which normally have only a single superpeer connection to its parent.

Finally, new peers entering the network will "mirror" the mode of operation of their connection point: a new node will run in either the normal Myconet mode (in which case it will start in the biomass state), or bulwark mode. The new peer will also mirror the concentration of alert hormone of the first peer it gets attached to; this ensures that the new peers have a head start on assessing the current state of the network without, and that they do not act as a net drain on the levels of alert hormone already in the system.

Signalling Overhead. During normal Myconet operations, peers maintain knowledge about their neighbors, and periodically signal changes to their protocol state, capacity, and neighbor count. The hormone diffusion signals discussed in this work are be piggybacked onto these updates at the cost of only a few bytes increase in message size, and no extra messages with respect to regular Myconet operation. We point the interested reader to [8], for a discussion of the typical overhead Myconet requires to maintain its overlay topology.

2.3   Parameters

The attack detection mechanism described in this paper has a number of tunable parameters. A discussion of how we approached finding values for these paramters and the effects of varying them can be found in Section 4.1.

Threshold concentrations of alert hormone (b_{enter} and b_{revert}) cause peers to enter and return from the bulwark state, respectively. These can be considered a "reference" parameters against which the rest hormone parameters can be calibrated. While these thresholds can be different, they were set to be the same value in our experimental work: $b_{enter} = b_{revert} = 50$.

Coefficients of the hormone generation formula (F_Q, F_L, and F_S), such that when a node of degree d fails, its former neighbors generate $F_Q d^2 + F_L d + F_S$. For these experiments, we used $F_Q = 0.5$, $F_l = -2$, and $F_S = 0$.

Hormone decay rate (h_{decay}) controls how quickly each peer metabolizes the hormone and "forgets" about damage and failures. For the current work, $h_{decay} = 0.15$ was used as the reference value.

Maximum local hormone concentration (h_{max}) caps the amount of hormone that can accumulate in each peer. Notice that this capping only occurs after other hormone diffusion dynamics have taken place, so large spikes cause by high-degree peers will be allowed to diffuse throughout the neighborhood. For this paper, $h_{max} = 250$ was used.

Inhibition delay (i_{enter} and i_{revert}) for switches to and from the bulwark state. Our current implementation is built on top of a cycle-based simulation, so these parameters represent the number of cycles that a peer must be above the threshold b_{enter} (in the case of i_{enter}) or below b_{revert} (in the case of i_{revert}). Values of $i_{enter} = 1$ and $i_{revert} = 5$ were used.

Minimum and maximum number of neighbors maintained by bulwark nodes (b_{min} and b_{max}) characterize the targeted density of the network when in the flattened state. Bulwark nodes will attempt to connect to at least b_{min} neighbors, but will permit as many as b_{max}, should other nodes connect to it through random dynamics. If a bulwark peer is over degree b_{max}, it will select neighbors at random to drop. Values of $b_{min} = 4$ and $b_{max} = 6$ were used.

3 Evaluation

3.1 Experiment Design

In order to test its performance in detecting and thwarting attacks, we implemented our hormone–based strategy on top of an existing cycle-based simulation of the Myconet protocol (described in [7,8]). This simulation uses the Java-based PeerSim framework [12], a widely-used platform for evaluating peer-to-peer protocols. In addition to the parameters of our hormone-based strategy, which we discussed in section 2.3, the Myconet protocol itself has a number of customizable parameters that are relevant to our experiments.

C_n represents the target number of inter-hyphal links that branching and immobile nodes maintain, and impacts the level of resilience of the Myconet overlay to disruptive events. For all experiments, we used $C_n = 5$; that value has been validated empirically in our previous work as providing a desirable amount of resilience. Also, Myconet peer capacities are assigned using a power-law distribution, but are capped to a maximum value c_{max}. Of course, determining an appropriate value for c_{max} may depend on the application. We chose $c_{max} = 500$ for experiments with networks of 10,000 peers, as this value is used for evaluating similarly scaled protocols in [2] and [6]. Since very large values of c_{max} relative to the number of peers in the network tend to produce degenerated superpeer topologies, we set $c_{max} = 500$ for network sizes between 1,000 and 10,000 nodes, and $c_{max} = 50$ for experiments of 1,000 nodes and smaller.

In our experiments, we simulate denial-of-service against the most important peers in the network, similar to the attack described in [10]: while an attack is in progress, each cycle the k highest-degree peers are removed (for all results in this section, $k = 2$ was used). Attacks begin at round 40 (after the superpeer topology has been constructed by the Myconet protocol and has stabilized, with easily recognizable, prominent hyphal peers) and continue for ten cycles. Each experiment discussed in this section, consists of 100 independent runs, each lasting 100 cycles; we report the average of a set of metrics over those 100 runs.

In addition to metrics examining the topological differences of the Myconet overlay when switching between the superpeer and the flat mode, we present metrics that evaluate the following properties:

- **Attack detection accuracy**, that is, whether the network recognizes that an attack is occurring, and successfully switches mode (when the number of peers in the bulwark state outnumbers the peers in any of the other Myconet protocol states). False negatives are counted when the network fails to switch mode for the whole attack period; false positives occur if the network switches mode outside of an attack period, for example, because of node failures that are induced by regular network churn. Notice that we can encounter both false positives and false negatives in the same experiment.
- **Attack detection time**, which is expressed as the number of cycles needed to perform a mode switch from the moment an attack begins, plus the number of cycles needed for all of the remaining peers to switch into the bulwark state.
- **Reversion latency**, the amount of time for the peers to switch back from the bulwark state into normal operation. As reorganizing the network topology is an expensive operation, the system should not switch back too quickly. For this purpose, the reversion inhibition parameter (i_{revert}) is used to delay the return.

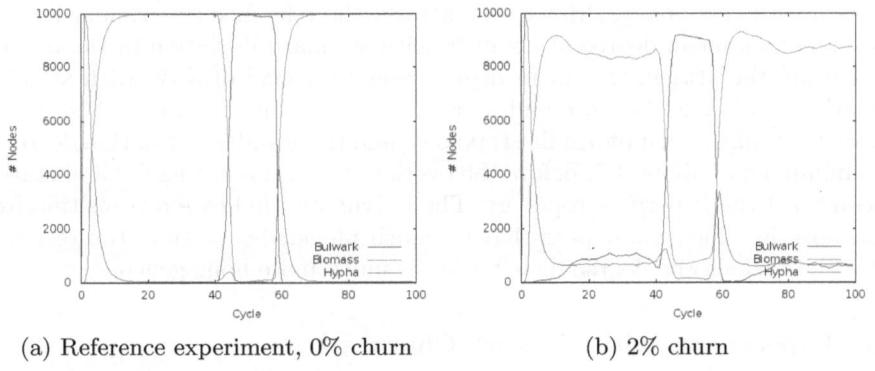

(a) Reference experiment, 0% churn   (b) 2% churn

Fig. 3. Mean nodes in different protocol states (10,000 nodes, 10-round attack beginning at cycle 40)

3.2   Reference Experiment

First of all, we conducted an experiment with a *reference network configuration*, with respect to the major parameters discussed above. The reference experiment provides us with a performance baseline, and allows us to gauge the quality of our attack-countering strategy. We then conducted other experiments, varying some of the parameters, to study the sensitivity of our strategy to different conditions.

For our reference experiment we use a network with 10,000 peers, $c_{max} = 500$, $C_n = 5$, no churn, and the rate of hormone decay set at $h_{decay} = 0.15$.

A single attack is started at round 40, which removes the two largest-degree nodes each cycle for ten cycles. Under these settings, attacks are detected with 100% accuracy (no false negatives), within a mean 4.59 cycles, and all nodes in the network have switched over to the bulwark state in a mean 7.55 cycles. Following the end of the attack, the system lingers in the bulwark state for a mean 9.92 rounds before at least 50% of the nodes have reverted to normal Myconet operation. This reversion latency includes a 5-rounds additional delay induced by the i_{revert} parameter, during which a peer lingers in the bulwark state, after its hormone concentration has dropped below the b_{revert} threshold.

To illustrate the behavior of the peer-to-peer network under attack, we examine the dynamics of how peers change state: the chart in Figure 3a plots the mean number of peers in either the biomass, hyphal (i.e., superpeer), or bulwark states over time, before, during and after an attack. All nodes start in the biomass state, then the hypha count spikes as the network bootstraps itself, which is a typical feature of the Myconet protocol. Then, the number of hyphae quickly drops as Myconet identifies the best superpeers and most of the others revert to biomass state. As the attack begins, the number of bulwark peers rises, initially slowly; then, remainder of the peers follow suit, until the network as a whole is in bulwark state. At that point, the mode switch is complete, and the network has effectively transitioned to a flat topology with no recognizable superpeers.

To verify that, we can examine the mean node degree of peers in the different phases mentioned above. Prior to the attack, the reference experiment network converges to a mean degree of about 2, with standard deviation in the order of 20. During the attack, the mean degree rises to a max of 5.12, with standard deviation at of 7.22. As more and more nodes move into the bulwark state, the network establishes a uniform flat topology, and the standard deviation decreases to a minimum of about 4.5, before the overlay switches mode again, and start to re-construct the superpeer topology. These dynamics indicate a transition from a network in which a few peers have a much higher degree than the rest, to a network in which the degree distribution is much more homogeneous.

3.3 Experiments with Network Churn

We also examined the performance of our technique when peer churn is constantly present in the P2P network. We used a simple model of churn, which replaces a fixed percentage of the peers each round, selected at random from the whole population. We tested churn levels from 1% to 3% (that is, for a 10,000-peer network, with churn varying from 100 to 300 nodes per cycle). In order to keep the capacity distribution the same over the length of a run, each removed node is replaced with a new, disconnected node of equal capacity.

Performance for 0%, 1%, 2%, and 3% churn is shown in the rightmost column of Table 1. As it can be expected, increasing levels of churn causes an increase of false positives. As a consequence of these random node failures, some "background level" of alert hormone is present in the network at all times. As shown in Figure 3b, under churn, a non-zero percentage of nodes is likely to be in the bulwark state at all times. If the churn happens to hit a number of high-degree nodes in close succession, it is possible to witness a false positive, that is, a switch of the network into the flat topology in absence of an actual targeted attack.

False negatives also increase under churn, e.g., when a portion of the network has switched into the bulwark state and a sizable proportion of peers are below the reversion threshold, but still within the delay window induced by the i_{revert} parameter. Since nodes in bulwark state are inhibited from emitting alert hormone, this may sometimes impede hormone diffusion while the attack progresses, preventing the cascade of the remaining peers into the bulwark state.

It is noticeable that the average attack detection times and time to switch the full network into bulwark were somewhat faster in experiments with churn (3.97 and 6.66 cycles for 1%, 3.72 and 6.64 for 2%, 3.96 and 7.06 for 3%, vs. 4.59 and 7.55 in the stable network). Similarly, time to revert to normal operation after an attack was also slightly faster (9.51 cycles for 1%, 9.19 for 2%, and 9.18 for 3%, vs 9.92 for 0%).

3.4   Experiments Varying Network Size

Varying the number of nodes in the simulation provides additional insight into the alert hormone dynamics. In addition to the reference experiments with 10,000 peers, additional experiments were run with 1,000 peers (with $c_{max} = 50$) and 5,000 peers (with $c_{max} = 500$). As can be seen in Table 1, using the reference parameters and a rate of hormone decay set at 0.15, a 1,000-peer network may miss up to about half of the attacks in the 0% churn case, whereas performance under churn is much better. The 5,000-node network shows the opposite behavior, with all attacks detected in case of 0% churn, but with a rise in both false positives (and false negatives) as the churn level increases.

The reason for those behaviors is illuminated by varying the decay rate, which alters the background levels of alert hormone present in the system under churn. On the one hand, as shown with $h_{decay} = 0.18$ and 5,000 peers, faster decay induces better performance under churn. On the other hand, fast decay rates lower the total amount of hormone in circulation in the network. In a smaller network (as in the 1,000 peer case with $h_{decay} = 0.15$), coupled with the absence of a constant level of hormone generated by churn, the result is significant number of false negatives, as it is harder to reach suficient concentrations of hormones in enough nodes to trigger a bulwark cascade. By tuning h_{decay} to 0.12 we can greatly lower the level of false negatives for 0% churn in relatively small networks, to the expense of more false positives at higher churn levels.

Interestingly, variation in the number of peers and h_{decay} also changes the speed at which the system switches state. For a configuration with 1,000 nodes, 0% churn and $h_{decay} = 0.15$, the system has a mean detection speed of 5.90 cycles and a reversion latency of 9.82 cycles. Changing the decay rate to 0.12 raises these numbers to 7.95 and 12.49 cycles. For 5,000 nodes, 0% churn and $h_{decay} = 0.15$, mean detection speed is 3.89 and mean reversion latency is 9.03. Raising the decay rate to 0.18 results in faster detection speed (3.76 cycles) and reversion speed (7.71 cycles).

4   Discussion

4.1   Parameter Adjustment and Sensitivity

When designing a self-organizing algorithm, a significant portion of the work revolves around designing a parameter space and choosing appropriate values that support the desired convergent behavior. There are no magic numbers, though; values may not be absolute, and may depend strongly on other properties. In the case of the attack detection and mitigarion strategy discussed in this paper, factors such as the environmental conditions (e.g., churn levels) and characteristics of the peers and the network as a whole have significant effects.

Table 1. False negative (FN) and false positive (FP) rates for attack detection (reference experiment in gray)

Nodes	1000		5000		10000
h_{decay}	0.15	0.12	0.15	0.18	0.15
Churn	FN FP	FN FP	FN FP	FN FP	FN FP
0%	0.51 0.00	0.06 0.00	0.00 0.00	0.00 0.00	0.00 0.00
1%	0.08 0.01	0.06 0.10	0.11 0.32	0.08 0.22	0.03 0.10
2%	0.01 0.11	0.04 0.30	0.18 0.50	0.08 0.34	0.10 0.16
3%	0.06 0.23	0.06 0.38	0.21 0.50	0.10 0.34	0.19 0.24

Among the critical parameter choices are the hormone production coefficients F_Q and F_L: if they are too high, the effect of even minor failures is magnified. In early experiments, a negative F_S was used to reduce the effect of small node failures, but this effect can also be achieved with a negative F_L.

The h_{decay} is also very important. A low value will cause hormone levels to slowly ramp up even in the face of normal churn, eventually forcing peer to switch to bulwark (and potentially staying in that state for longer than desired). A high value dissipates the hormone too quickly, resulting in failure to detect even large attacks. The correct trade-off, as we have already discussed in our evaluation, is not a constant, but – all other considerations being equal – a function of the expected number of peers, peer degree distributions, and churn dynamics.

If h_{max} is very large, it will allow very large reservoirs of hormone to develop, resulting in an overlay that takes a very long time to revert to the superpeer mode. If h_{max} is too small, it may mute the effect of node failure, preventing sufficient quantities of hormone from pushing nodes past their alert thresholds.

i_{enter} prevents local spikes from instantly causing nodes to jump into bulwark, allowing time for diffusion (with the possibility of new infusions of alert hormone from other nodes). Similarly, once the network has switched over to bulwark operations, the generation of new alert hormone falls off, and the reversion delay provides a threshold of pessimism, ensuring that the cost of switching the network state does not have to be incurred again too quickly. A value of $i_{enter} = 1$ was found to eliminate much of the tendency of lone nodes to jump into bulwark before other nodes in the neighborhood were also close to ready. $i_{revert} = 5$ was used to cause nodes to "pessimistically linger" in bulwark for an additional five rounds after they had fallen below the threshold.

4.2 Future Work

For all the parameters discussed above, our choices were hand-selected based on observation of their effects during simulation. A natural next step is to use a learning mechanism (such as a genetic algorithm) to explore the parameter space and find optimized combinations of parameters, leading to more accurate detection of failures. If the learning is performed online, that can enable another level of self-adaptation, allowing the system to respond to changing conditions that had not been foreseen during design.

Various measures of network robustness have been proposed, e.g., in [13,14,15]. An analysis of the networks resulting from the use of the alert hormone mechanism proposed in this paper is another direction of investigation. While instant resistance to percolation may not capture the dynamic robustness properties of self-healing networks, variants of these metrics may be useful in capturing the differences between normal operations and the flat mode.

The attack detection approach in this paper focuses on defending from "crash attacks", and does not address sneakier malicious behavior involving rogue peers. There is a large literature on attacks against P2P networks (see, e.g., citeabdelouahab2009tcat). In fact, it is known that using self-organization or online learning techniques can make a system vulnerable in additional ways, to attackers that are aware of the logic that adapts the system behavior. For example [16] dsicussed the "frog-boiling" attack, where malicious peers slowly push a system into a degraded state while remaining within "expected" thresholds of behavior.

Another direction for future work to examine potential attacks against the diffusive attack detection mechanism, in particular "false positive" attacks where attackers may achieve disruption of the network operations by devising ways to generate a false or disproportionate influx of alert hormone. Heuristics or learning mechanisms may help mitigate this type of attack.

5 Related Work

There is a vast literature on attacks on P2P networks and methods to thwart them. Yue et al. [17] offered a taxonomy of attacks, including various types of identity assignment attacks, attacks against the routing process, and application attacks. In this work, we focus on "crash" attacks occurring at the overlay level: attacks that aim at disconnecting or seriously disrupting the fabric of the P2P network by disabling or eliminating some of the peers.

Topology disruptions are a threat for both structured and unstructured P2P overlays. In structured overlays, it is crucial to maintain a specific topology, because the properties of the application running on top of the overlay (e.g., routing and lookup in a DHT) cannot be guaranteed otherwise. Thus, defenses that themselves alter the topology of the network are not very common. Instead, structured overlay protocols typically employ *topology preservation* defenses, that is, maintain a built-in degree of redundancy so that alternate paths to resources are available if some nodes fail or are otherwise removed. Such a property, although not sufficient per se in case of Byzantine threats, significantly reduces the threat of crash attacks [18]. A kind of topology adaptation defense is offered by [19], which discusses how the Tapestry structured overlay may identify and neutralize Denial-Of-Service (DOS) attacks by isolating attacking peers in ad hoc clusters.

Unstructured overlays do not have the same, strict topology requirements; moreover, when they organize into a scale-free topology, they give up redundancy as a defense against crash attacks against superpeers. For this reason, *topology recovery* defenses are common in unstructured overlays. The focus is often on protocols (such as SG-1 [2] or ERASP [6]) that can heal the fabric of the P2P network, whenever it is disrupted by a successful targeted attack. This is a kind of reactive defense; it minimizes damage to the P2P network and the applications it hosts, but cannot, of course, thwart the attack itself.

Topology adaptation defenses are more proactive, and could be used to discourage or counter attacks by making them ineffective or costly. However, topology adaptation is itself a cost: [20] examines the effect of failures of nodes on unstructured P2P networks, and analyzes the efficacy of increasing redundancy, degree of connectivity, and hop-count distance traversed by search requests.

The goal of our topology adaptation defense is eliminating the superpeers as weak points when they are threatened. Our direct inspiration comes from the works of Zweig and Zimmerman [10], and Keyani *et al.* [9]. The former offers a graph-theoretical argument showing the feasibility and effectiveness of progressively flattening the range of node degrees in an overlay in response to attacks, but it does not report empirical results for comparison.

The latter discusses an implementation of a defense that switches a Gnutella overlay from a scale-free to an exponential topology, therefore it represents the closest approach to which we can compare. This approach is able to adapt the topology to protect against attack based on detection occuring in a relatively small proportion of the nodes (less than 15%). However, it does not discuss a mechanism for reverting the topology after an attack is complete, the initial scale-free topology is built using a centralized oracle. In our work, we demonstrate how both mode changes can occur in a fully decentralized way (thanks to the dynamics of hormone diffusion) which is important because, in the absence of an attack, the superpeer topology provides superior efficiency in exploiting the natural heterogeneity of node capabilities in peer-to-peer applicaitons, as well as strong robustness to random failure. The diffusive approach described in this paper also allows all nodes to detect that an attack is underway, allowing all nodes to participate in the topology adaptation.

One characteristic of our work is the usage of hormone diffusion as the self-organized principle that ensures the convergence of the behavior of individual peers. One of the earliest papers discussing diffusion as basis for self-organization is [21], where Turing described the reaction-diffusion model. Contemporary diffusion models that pertain to the management and adaptation of computing networks include [22] and [23], which use hormones to manage the request load in a network. Works on gradient routing [24] and directed diffusion [25] use instead a diffusive mechanism for resource location in a network.

To the best of our knowledge, our work is the first that employs diffusive signalling as the basis of a P2P defense strategy.

6   Conclusions

We have presented a biologically–inspired approach for defending an unstructured P2P network from attacks that aim at disrupting the overlay by targeting superpeers. Our approach is fully self-organized and is based on a diffusion mechanism that propagates an alert hormone among peers.

The concentration of the hormone within each individual peer determines whether that peer uses protocol rules that build either a superpeer-based topology, or a flat topology with a narrow degree distribution. The dynamics of hormone diffusion, concentration and metabolization can be adjusted to ensure that the network as a whole distinguishes attacks from regular peer churn, and transitions quickly and accurately between those two topologies.

Our approach shows that a topology switch defense against these targeted attacks can be effective and easy to implement.

References

1. Yang, B., Garcia-Molina, H.: Designing a super-peer network. In: Proceedings of the 19th International Conference on Data Engineering (ICDE), p. 49 (2003)
2. Montresor, A.: A robust protocol for building superpeer overlay topologies. In: Proceedings of the 4th International Conference on Peer-to-Peer Computing, Zurich, Switzerland, pp. 202–209. IEEE (August 2004)
3. Gnutella Protocol Specification 0.6, http://rfc-gnutella.sourceforge.net
4. Liang, J., Kumar, R., Ross, K.W.: Understanding KaZaA. Polytechnic Univ. (2004) (manuscript)
5. Baset, S.A., Schulzrinne, H.: An analysis of the skype peer-to-peer internet telephony protocol. In: Proceedings IEEE INFOCOM 2006 25th IEEE International Conference on Computer Communications, vol. 6, pp. 1–11. IEEE (2004)
6. Liu, W., Yu, J., Song, J., Lan, X., Cao, B.: ERASP: An Efficient and Robust Adaptive Superpeer Overlay Network. In: Zhang, Y., Yu, G., Bertino, E., Xu, G. (eds.) APWeb 2008. LNCS, vol. 4976, pp. 468–474. Springer, Heidelberg (2008)
7. Snyder, P.L., Greenstadt, R., Valetto, G.: Myconet: A fungi-inspired model for superpeer-based peer-to-peer overlay topologies. In: SASO 2009, pp. 40–50 (2009)
8. Valetto, G., Snyder, P.L., Dubuois, D.J., Di Nitto, E., Calcavecchia, N.M.: A self-organized load-balancing algorithm for overlay-based decentralized service networks. In: SASO 2011 (2011)
9. Keyani, P., Larson, B., Senthil, M.: Peer pressure: Distributed Recovery from Attacks in Peer-to-Peer Systems. In: Gregori, E., Cherkasova, L., Cugola, G., Panzieri, F., Picco, G.P. (eds.) NETWORKING 2002. LNCS, vol. 2376, pp. 306–320. Springer, Heidelberg (2002)
10. Zweig, K.A., Zimmermann, K.: Wanderer between the worlds - self-organized network stability in attack and random failure scenarios. In: Second IEEE International Conference on Self-Adaptive and Self-Organizing Systems, SASO 2008, pp. 309–318. IEEE (2008)

11. Jelasity, M., Kowalczyk, W., Van Steen, M.: Newscast computing. Vrije Universiteit Amsterdam Department of Comp. Sci. Internal Report IR-CS-006 (2003)
12. Jelasity, M., Montresor, A., Jesi, G.P., Voulgaris, S.: The Peersim simulator, http://peersim.sf.net
13. Albert, R., Jeong, H., Barabási, A.L.: Error and attack tolerance of complex networks. Nature 406(6794), 378–382 (2000)
14. Van Mieghem, P., Doerr, C., Wang, H., Hernandez, J.M., Hutchison, D., Karaliopoulos, M., Kooij, R.E.: A framework for computing topological network robustness. Delft University of Technology, Report20101218 (2010), www.nas.ewi.tudelft.nl/people/Piet/TUDelftReports
15. Ghedini, C.G., Ribeiro, C.H.C.: Rethinking failure and attack tolerance assessment in complex networks. Physica A: Statistical Mechanics and its Applications (2011)
16. Chan-Tin, E., Feldman, D., Hopper, N., Kim, Y.: The Frog-Boiling Attack: Limitations of Anomaly Detection for Secure Network Coordinate Systems. In: Chen, Y., Dimitriou, T.D., Zhou, J. (eds.) SecureComm 2009. LNICST, vol. 19, pp. 448–458. Springer, Heidelberg (2009)
17. Yue, X., Qiu, X., Ji, Y., Zhang, C.: P2p attack taxonomy and relationship analysis. In: 11th International Conference on Advanced Communication Technology, ICACT 2009, vol. 02, pp. 1207–1210 (February 2009)
18. Srivatsa, M., Liu, L.: Vulnerabilities and security threats in structured overlay networks: A quantitative analysis. In: Computer Security Applications Conference, Annual, pp. 252–261 (2004)
19. Perlegos, P.: Dos defense in structured peer-to-peer networks. Computer (2004)
20. Samant, K., Bhattacharyya, S.: Topology, search, and fault tolerance in unstructured p2p networks (2004)
21. Turing, A.M.: The chemical basis of morphogenesis 237(641), 37–72 (1952)
22. Balasubramaniam, S., Botvich, D., Donnelly, W., Foghlú, M.Ó., Strassner, J.: Biologically inspired self-governance and self-organisation for autonomic networks. In: Proceedings of the 1st International Conference on Bio Inspired Models of Network, Information and Computing Systems, BIONETICS 2006. ACM, New York (2006)
23. Trumler, W., Thiemann, T., Ungerer, T.: An artificial hormone system for self-organization of networked nodes (2006)
24. Poor, R.D.: Embedded Networks: Pervasive, Low-Power, Wireless Connectivity. PhD thesis, Massachusetts Institute of Technology, School of Architecture and Planning, Program in Media Arts and Sciences (2001)
25. Intanagonwiwat, C., Govindan, R., Estrin, D.: Directed diffusion: A scalable and robust communication paradigm for sensor networks. In: Proceedings of the 6th Annual International Conference on Mobile Computing and Networking, pp. 56–67. ACM (2000)

Dynamic Hawk and Dove Games
within Flocks of Birds

Eitan Altman[1], Julien Gaillard[1,2], Majed Haddad[1,2], and Piotr Wiecek[3]

[1] INRIA Sophia Antipolis, 10 route des Lucioles, 06902 Sophia Antipolis, France
[2] LIA, University of Avignon, France
[3] Institute of Mathematics and Computer Science,
Wroclaw University of Technology, Wroclaw, Poland

Abstract. The Hawk and Dove game is a well known model from biology for competition over resources between two types of behaviors: aggressive (Hawk) and peaceful (Doves). The game allows to predict whether one of the behaviors will dominate the other or whether we may expect coexistence of both at a long run; in the latter case it allows to predict what fraction of the population will be aggressive and what peaceful. This game is quite relevant to networking, and has been used in the past to predict the outcome of competition between congestion [2] control protocols (both in wireline and in wireless) as well as between power control protocols for wireless communications. In this paper we study new aspects of the game within the framework of flocks of birds, and obtain results that can be useful for network engineering applications as well.

1 Introduction

Much of the behavior of flocks of birds is known to be related to very simple rules that determine the behavior of each individual in the flock. One of the rules concerns a repulsive force that acts on a bird whenever its territory is invaded by some other bird. The territory is defined as some ball around the bird. We consider the radius of this ball as measure of aggressiveness of a bird. We consider in this paper a variant of the well known Hawk and Dove game, in which the birds of the flock compete over resources have to determine whether to adopt an aggressive (Hawk, corresponding to a small radius) or a peaceful (Dove, large radius) behavior. As in the classical game, introduced by J. Meynard Smith [9], there is some advantage of adopting a Hawk behavior in that whenever the adversary is a Dove, the Hawk gets the resource. On the other hand, there is a disadvantage in being a Hawk, which is risking to get hurt, whenever the adversary is a Hawk.

We first study the mapping of this problem to the classical HD game. We then study a state dependent version of this problem in which each individual has an energy state. The result of contests over food determines the evolution of this state of each individual. An individual dies when its state reaches zero. Each individual seeks to maximize the sum of fitness over its life time. Our objective

E. Hart et al. (Eds.): BIONETICS 2011, LNICST 103, pp. 115–124, 2012.
© Institute for Computer Sciences, Social Informatics and Telecommunications Engineering 2012

is to determine whether an ESS (Evolutionary Stable Equilibrium) exists, i.e. whether at a long run we may expect one type of behavior to dominate or whether we may expect the two types to coexist; in the latter case we are interested in determining the fraction of the population that is aggressive at equilibrium. We study the situation in which each individual has an identity - Hawk or Dove, and thus the choice of being one or the other does not change during the lifetime of an individual. This is in contrast with previous work on state dependent HD [3,1,5] game in which an individual can make different choices at different energy states.

The state dependent Hawk and Dove game has been applied to power control problems in wireless in which the state represents the energy level of the battery and state zero represents an empty battery. This application is however different than the one we introduce here: it has the simpler structure in which a player's decision (whether to transmit at a high or at a low power) has an impact on the success of the current transmission, and on its own state evolution, but *not on the evolution of the state of other players*. Stochastic (state dependent) games in which a player does not have an influence on the state transition or other players are known to be much simpler than the general stochastic games [13]. We provide an explicit expression for the equilibrium behavior of the game.

The structure of the paper is as follows. In the next Section, we present the system model. In the Section 3, we describe the evolutionary game and address the properties of the fitness. In Section 4, we investigate the existence of ESS. We give explicit results on conditions for having pure or mixed equilibria. We conclude the paper in Section 5.

2   Model

We consider the following Markov Decision Evolutionary Game (a framework for state dependent evolutionary games introduced in [3,5,1]): There is a large population of individuals, say birds. A bird i is born at some time $T_0(n)$. Then it finds itself involved at some times $T_i(n)$ in a competition with another bird (each time it is another bird it competes with). A Hawk-Dove game is used to predict the behavior of the competing birds. We consider that an individual starts with some energy level n. If it looses contests over food then it becomes weaker: its energy level decreases. If it wins then the level stays the same. If the energy level reaches zero then the individual is assumed to die. We consider a variant of the well known Hawk and Dove game, in which the birds of the flock compete over resources have to determine whether to adopt an aggressive (Hawk) or a peaceful (Dove) behavior. We refer to these by hawks (H) and doves (D), respectively, thereby referring to the HD game. Considering only pairwise interaction, it is assumed that the sequence of types of individual with which a given individual interacts constitutes a sequence of i.i.d. random variables.

Transition probabilities. We assume that a fight between two H results in none having any offsprings in that period. For any encounter resulting in no fight (DD, DH or HD) the distribution of the number of offspring is the same. We assume that the additional number of offspring from a period without a fight is equal to 1. Further, we assume that natural selection occurs in a way that favors behavior that maximizes the total expected fitness during the lifetime of an individual. We also assume that the frequency of encounters does not depend on the energy level.

Aging factor. Life is bounded due to two features: energy and aging. The aging is modeled through an aging factor δ which does not depend on the type of the individual (H or D). It plays the role of a "discount factor" in optimal control. In absence of energy constraints, the life time of an individual is geometrically distributed with parameter δ.

Initial energy level. In the remainder, we assume that an individual starts at energy level (state) N_D or N_H, depending on the behavior of the individual (H or D). It represents the number of opportunities an individual can encounter during its lifetime. When the state equals zero, it dies. At that moment it is immediately replaced by some new individual.

3 Properties of the Fitness

Both hawks and doves aim to optimize the number of offspring during their lifetime, hence the fitness is defined as follows

Definition 1. *The long term fitness of an individual is defined as the total expected number of offspring of an individual that starts with energy level n during its lifetime. We denote by $V(j,i)$ the long term fitness of an individual, given that it is of type j, and that all others are of type i, with $i,j \in \{H, D\}$.*

Definition 2. *Assume that at any time, a fraction α of the individuals use action H, and the rest use D. We then denote by $V(j, \alpha)$ the corresponding long term fitness given that the individual uses j. Moreover, let*

$$V(\beta, \alpha) = \beta V(H, \alpha) + (1 - \beta)V(D, \alpha)$$

be the fitness of an individual that behaves as a H with probability β, and otherwise behaves as a D (with probability $1 - \beta$).

3.1 Definition of a Standard Evolutionary Game

Suppose that the whole population uses a strategy q and that a small fraction ϵ (called "mutations") adopts another strategy p. Evolutionary forces are expected to select against p if

$$V(q, \epsilon p + (1 - \epsilon)q) > V(p, \epsilon p + (1 - \epsilon)q) \tag{1}$$

Definition 3. *A strategy q is said to be an Evolutionary Stable Strategy (ESS) if for every $p \neq q$ there exists some $\epsilon_y > 0$ such that (1) holds for all $\epsilon \in (0, \bar{\epsilon}_y)$.*

We shall make use of the following characterization of an ESS [7]:

Theorem 3.1. *A strategy q is an Evolutionary Stable Strategy if and only if* $\forall p \neq q$ *the following conditions holds:*

$$V(q,q) \geq V(p,q), \tag{2}$$

and if

$$V(q,q) = V(p,q) \text{ then } V(q,p) > V(p,p). \tag{3}$$

The first condition says that the ESS is a Nash equilibrium in the game that describes the interaction between two players. Conversely, if q is a strict Nash equilibrium in that game then it is an ESS in the evolutionary game.

The second condition, referred to as "Maynard Smith's second condition", states that if q is a Nash equilibrium but not a strict Nash equilibrium (i.e. the fitness of a deviation p from q does as good as q when the rest of the population uses q), then q can still be an ESS if it has an advantage in that it can invade the mutants strategy p. In other words, in a population where every one uses p, a small deviation to q does strictly better than everyone using p.

We shall consider evolutionary games where each player has a finite number of available pure actions and where the set of strategies of a player is the set of probability distributions over his actions. Let $V(p,q)$ denote the expected fitness (utility) for a player when playing a mixed policy p and when the fraction of the population that plays each pure strategy i is given by $q(i)$. The expected fitness is then linear in both p and q and can be written as pVq^T where V is the matrix whose i,jth entry equals $V(i,j)$, and where p (resp. q) is a row vector whose ith entry is $p(i)$ (resp. $q(i)$). Theorem 3.1 then states that the ESS of an evolutionary game can be characterized by properties of the equilibria of an auxiliary game. In our case this auxiliary game is the matrix game V. Note that not every matrix game has an ESS.

3.2   ESS in the Dynamic Game

When writing the long term fitness of players as a function of the system parameters, we shall see that the fitness is linear in p and q whereby p are now probabilities over the strategies H and D and not over the actions H and D. This means that a mixed strategy is obtained by tossing a coin, and according to the outcome, the player always uses D or always uses H. Notice that if we choose between action D and H with some probability q at each time instant, then the expected fitness need not be linear in q. This bilinear form of dynamic games allows us to apply directly the standard theory of evolutionary games to the dynamic case. Recall that, even though we assume that each individual j always plays the same action, the sequence of actions that are played by those encountered by some tagged individual are i.i.d. While working with mixed strategies allows for directly applying much of the framework of standard evolutionary games, these policies do not allow for an evolution, as once we perform the initial randomized selection between D and H, we shall always stick to that choice. Hence, to combine both the flexibility that allows for evolution together with the linear properties of the auxiliary game (the matrix game introduced above), we assume that each individual uses mixed policies for some limited time, after which a new choice is made and so on. The definition of V (see Definition 2) is suitable for mixed strategies over an infinite time as well as for the finite horizon framework.

4 Computing the Equilibrium

Let $V_n(i, \alpha)$ denotes the expected fitness of an individual who plays i and starts at energy level n, $i, j \in \{H, D\}$. In view of this definition, we have $V(D, \alpha) = V_{N_D}(D, \alpha)$ and $V(H, \alpha) = V_{N_H}(H, \alpha)$. We find the following recursions for $V_n(i, \alpha)$,

$$V_n(H, \alpha) = \alpha \frac{(1-\delta)V_n + (1+\delta)V_{n-1}}{2} + (1-\alpha)(1 + (1-\delta)V_n + \delta V_{n-1}) \quad (4)$$

$$V_n(D, \alpha) = 1\{n > 0\} + \alpha V_{n-1} + (1-\alpha)\left(\frac{(1-\delta)V_n + (1+\delta)V_{n-1}}{2}\right) \quad (5)$$

Thus $V_n(\beta, \alpha)$ has the form

$$A V_n(\beta, \alpha) + B V_{n-1}(\beta, \alpha) = C \quad (6)$$

Equation (6) is a first order linear difference equation whose solution is

$$V_n(H, \alpha) = \frac{(1-\alpha)}{\delta + \frac{\alpha(1-\delta)}{2}} n, \quad (7)$$

$$V_n(D, \alpha) = \frac{1}{\alpha + \frac{(1-\alpha)(1+\delta)}{2}} n, \qquad n > 0, \quad (8)$$

whereby we assumed $V_0(D, \alpha) = V_0(H, \alpha) = 0$. That is, no fitness can be collected if the life ends.

Lemma 4.1. *We have for $i \in \{D, H\}$,*

$$V(i, \alpha) = \alpha V(i, H) + (1-\alpha)V(i, D), \quad (9)$$

with,

$$V(D, D) = \frac{N_{II}}{\delta}, \qquad\qquad V(H, D) = \frac{2 N_D}{1+\delta},$$

$$V(H, H) = 0, \qquad\qquad V(D, H) = N_D, \qquad\qquad \diamond$$

Notice that the behavior of the system depends on the ratio $\frac{N_D}{N_H}$ and not on N_H and N_D themselves. Let $\gamma = \frac{N_D}{N_H}$. This allows us to express the equilibrium as follows.

Corollary 4.1. *Fitness equations imply that:*

(i)   *H can not be a pure ESS since* $V(H,H) < V(D,H)$.

(ii)   *D is a pure ESS if* $\delta \geq \dfrac{1}{2\gamma - 1}$.

(iii)   *By setting* $V_{N_H}(H, \alpha) = V_{N_D}(D, \alpha)$, *we obtain the following quadratic equation in* α

$$\frac{1}{2}(\delta - 1)\alpha^2 + (1\delta - \frac{\gamma}{2}(1 - \delta))\alpha + \frac{1}{2}(1 + \delta) - \gamma\delta = 0$$

The solutions $\alpha_{1,2}^*$ *are expressed as*

$$\alpha_1^* = -1/2 \frac{\gamma\delta + 2\delta - \gamma - \sqrt{\gamma^2\delta^2 + 12\gamma\delta^2 - 2\gamma^2\delta - 12\gamma\delta + \gamma^2 + 4}}{-1 + \delta} \quad (10)$$

$$\alpha_2^* = -1/2 \frac{\gamma\delta + 2\delta - \gamma + \sqrt{\gamma^2\delta^2 + 12\gamma\delta^2 - 2\gamma^2\delta - 12\gamma\delta + \gamma^2 + 4}}{-1 + \delta} \quad (11)$$

If α^* *is in the interior of the unit interval then it is a mixed ESS.* ◇

Conditions on the Existence of ESS Equilibria: Notice that the existence of the mixed ESS α^* is still not insured. Indeed, one must identify conditions on parameter δ and γ in order to guarantee that α^* ranges between 0 and 1. We have the following conditions to be satisfied

Proposition 1. *The dynamic game considered in the paper has a mixed ESS. Moreover, we have*

- α_1 *is an ESS iff* $\frac{3}{2} \leq \gamma < \frac{1+\delta}{2\delta}$
- α_2 *is an ESS iff* $\sup\{\frac{3}{2}, \frac{4\delta}{1+\delta}\} < \gamma \leq \frac{1+\delta}{2\delta}$.

Proof: We have to check that $\alpha_{1,2} \in [0,1]$ and find the condition on the square root. □

However, these conditions do not seem very restrictive, as it is rational to assume that the aging factor is relatively small – otherwise there would not be any difference between periods when an individual has and does not have food, which intuitively does not make sense. Notice that there exists a fixed threshold γ^* for which we have

- For $\gamma > \gamma^*$, α is a monotically decreasing function of δ. We have an ESS iff $\delta \leq \frac{\gamma}{4-\gamma}$.
- For $\gamma < \gamma^*$, α is a monotically increasing function of δ. We have an ESS iff $\delta \leq \frac{1}{2\gamma-1}$.

The existence of such a threshold is not surprising, since two factors affect the lifetime of an individual: its N_i and the amount of food it eats. Note now that aging coefficient δ implies the difference in aging speed between a hungry and a full individual – the bigger δ is, the smaller this difference is. Thus, as δ grows, the bigger the impact of the ratio γ is, and so, when it is small the fraction of hawks in the equilibrium population grows, while when it is big, the same fraction decreases. It is important to notice that the threshold $\gamma^* \approx 1.5 > 1$, which shows that in case when biological clock of a hawk beats at approximately the same speed as that of a dove, hawks have visible advantage over doves, which is close to the intuition that we have from the classical HD game.

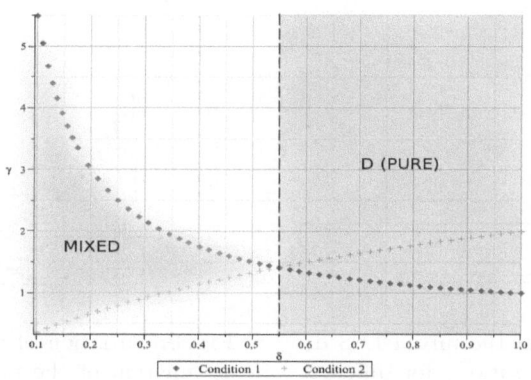

Fig. 1. γ as function of δ

In Figure 1, we clearly see that those conditions determine the mixed equilibrium region. **On the contrary, we see that there's no particular condition concerning the pure D equilibrium.**

5   Discussion

We first notice that H cannot be a pure ESS. This is in contrast with a standard HD game, where for some values of the penalty for fighting it is possible [11]. However, in our game we assume that the penalty for fighting is a lack of ability to have offspring in the period when fight takes place. Thus in a population where there are only hawks, no player can have any offspring at any time, while in a population with any small fraction of doves the fitness of both hawks and doves increases.

On the contrary, it is shown in Corollary 4.1 that D can be an ESS. However, as one can easily see, this is possible only if $\gamma \geq 1$. Of course, in the standard HD game, D is never an ESS, but in the $\gamma \geq 1$ case, doves take advantage of their prolonged lifetime. Note that in our setting we do not assume that the lack of food affects directly the number of offspring. The influence is indirect, through shorter lifetime, however $\gamma \geq 1$ means that the biological clock of a hawk beats faster than that of a dove, which mitigates the offspring-reducing effect of lack of food.

For any fixed value of δ, α_2^* decreases as a function of γ. This is easy to explain, as bigger γ means longer lifetime of a dove in comparison to that of a hawk. This obviously implies bigger number of offspring for doves and in consequence a bigger fraction of doves in equilibrium.

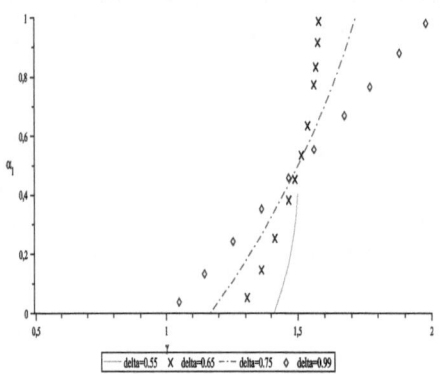

Fig. 2. Variation of the mixed ESS α_1 as function of the ratio γ for different aging factor δ

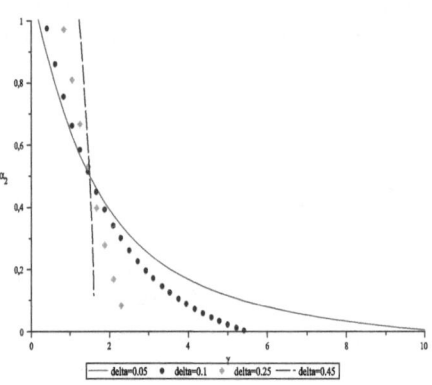

Fig. 3. Variation of the mixed ESS α_2 as function of the ratio γ for different aging factor δ

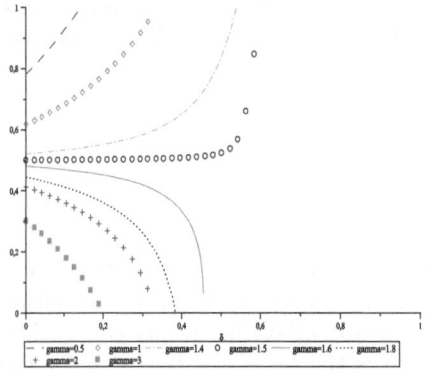

Fig. 4. Variation of the mixed ESS α_2 as function of the aging factor δ for different ratio γ

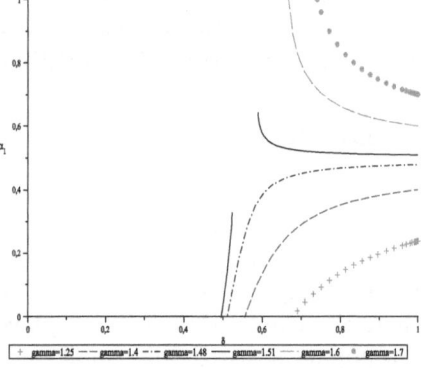

Fig. 5. Variation of the mixed ESS α_1 as function of the aging factor δ for different ratio γ

Figure 3 depicts the fact that as the aging factor increases, the ratio γ has more impact on the equilibrium. In other words, the aging factor influences the sensitivity of the equilibrium with respect to the ratio γ. It means that for a small aging factor δ, a change on gamma will not have a big impact on the equilibrium, whereas with a important aging factor δ, a small change in γ will have enormous consequences on the equilibrium. Hence, biologically speaking, a small mutation can modify considerably the equilibrium depending on the aging factor. Note that the tendency is the same for all δ.

On the contrary, Figure 2 shows a paradox in the way that increasing γ should give an advantage to the doves, and thus the number of doves at the equilibirum should increase too, but we observe the opposite phenomena.

Figure 4 shows the effect of the aging factor δ on the equilibrium for several values of the ratio γ. Here we can see that depending on γ, we have two different behaviors, separated by a threshold $\gamma^* \approx 1.5$, as mentioned before. In the case where $\gamma > \gamma^*$, as the aging factor increases, the proportion of hawks at equilibrium decreases, and this happens faster as γ increases. This is exactly the opposite for the case where $\gamma < \gamma^*$. This result confirms what we have seen before.

6   Conclusion

We have studied a new variant of the HD game in which the decision of an individual of being aggressive or peaceful is done not only for one given interaction with another individual, but for all future interactions that will occur during the its lifetime. We have derived explicit expressions for the equilibria and studied their properties. In particular, we note that in the original HD game (which is not state dependent), the pure policy H may be an equilibrium whereas D cannot. We notice that this is not what happens in the proposed dynamic HD game where H cannot be an equilibrium but D can be one.

References

1. Altman, E., Hayel, Y., Tembine, H., El-Azouzi, R.: Markov Decision Evolutionary Games with Expected Average Fitness. Evolutionary Ecology Research 11(4), 677–689 (2009)
2. Altman, E., El-Azouzi, R., Hayel, Y., Tembine, H.: The evolution of transport protocols: An evolutionary game perspective. Computer Networks 53(10), 1751–1759 (2009)
3. Altman, E., Hayel, Y.: Markov Decision Evolutionary Games. IEEE Transactions on Automatic Control 55(6) (June 2010)
4. Tembine, H., Altman, E., El-Azouzi, R.: Delayed Evolutionary Game Dynamics applied to the Medium Access Control. In: Bionetworks 2007, Pisa, Italy (2007)
5. Hayel, Y., Tembine, H., Altman, E., El-Azouzi, R.: A Markov Decision Evolutionary Game for Individual Energy Management. Annals of the International Society of Dynamic Games (2009)
6. McNamara, J., Merad, S., Collins, E.: The Hawk-Dove Game as an Average-Cost Problem. Avdances in Applied Probability 23(4), 667–682 (1991)
7. Maynard Smith, J.: Evolution and the Theory of Games. Cambridge University Press (1982)
8. MacNamara, J., Houston, A.: If animals know their own fighting ability, the evolutionary stable level of fighting is reduced. Journal of Theoretical Biology 232, 1–6 (2005)

9. Maynard Smith, J.: Game Theory and the Evolution of Fighting. In: Maynard Smith, J. (ed.) On Evolution, pp. 8–28. Edinburgh University Press, Edinburgh (1972)
10. Cressman, R.: Evolutionary Dynamics and Extensive Form Games. MIT Press (2003)
11. Berger, U.: Best response dynamics and Nash dynamics for games, Dissertation. University of Vienna (1998)
12. Hofbauer, J.: From Nash and Brown to Maynard Smith: Equilibria, dynamics and ESS. Selection 1, 81–88 (2000)
13. Filar, J., Vrieze, K.: Competitive Markov decision processes. Springer (1996)

Some Initial Experiments in Calibrating Emissions Models Using an Evolutionary Algorithm

Neil Urquhart

Centre for Emergent Computing, Edinburgh Napier University,
Edinburgh, United Kingdom

Abstract. Vehicle routing has traditionally been considered from an optimisation perspective in relation to minimising costs associated with distance travelled and number of vehicles used. Increasingly however, there is a demand to additionally optimise journeys according to the levels of carbon emissions. Incorporating this criterion into an optimisation algorithm necessitates the use of vehicle emission models. Although a number exist, they are often complex to use and customise due to the number of parameters involved. In this paper, we evaluate the use of an Evolutionary Algorithm to calibrate the parameters of a vehicle emissions model against real-world observed emissions data . The calibrated model can then be used with confidence within the fitness function of an optimisation technique on similar data. Initial results obtained suggest that this approach shows promise. The work forms the initial stages of a wider programme of work to investigate the use of nature inspired methods to construct accurate vehicle emissions models for use in green logistics.

Keywords: Evolutionary Algorithms, Optimisation, Carbon Reduction, Transport.

1 Introduction and Rationale

Increasing awareness of environmental issues within organisations and individuals is leading to a desire to minimise the environmental impact of their activities in areas such as vehicle routing and logistics. Nature inspired techniques have been applied to vehicle routing problems, such as the Vehicle Routing with Time Windows problem (VRPTW) with success for some time (see [1] for an overview). Traditionally, optimisation criteria include the number of vehicles utilised, distance covered and both fixed and variable costs relating to vehicle usage and fuel. Such quantities are generally straightforward to calculate, for example, distance can be calculated using a simple Euclidean function or more realistically using mapping data. Inclusion of CO_2 usage metric for evaluating a solution requires the use of a vehicle emission model. Emissions modelling (see section 2.1) is a complex area due to a number of factors such as differing road conditions, vehicle types and environmental factors which affect emissions (discussed later in section 3). Even when an appropriate model has been selected, it may be difficult to use due to the number of parameters utilised by the model. In this paper we discuss the calibration of an existing emissions model against actual emissions data. An evolutionary algorithm is used to evolve a set of parameters which represent the best-fit of the model to the data.

E. Hart et al. (Eds.): BIONETICS 2011, LNICST 103, pp. 125–132, 2012.

The proposed methodology will enable the model to be calibrated for use with a range of vehicle types, not just those for which existing parameter sets exist [1]. Having produced a calibrated model for a given vehicle type (in this case the Ford Mondeo vehicle used to generate the test data) the model may be used to predict emissions for that vehicle on a given drive cycle. Drive cycle data (e.g. latitude and longitude derived data) may be generated using GPS enabled mobile devices mounted on a vehicle. The training data utilised (see section 3) is based upon urban driving cycles within the United Kingdom, further testing with data from other geographical areas would be required to establish whether the models are applicable to other, similar, UK traffic situations. The primary area of application for these models would be in the area of micro simulation of road networks.

2   Modelling Vehicle Emissions

2.1   Emissions Models

A number of models have been developed to estimate the likely emissions of a vehicle in particular traffic conditions. Such models range in granularity from full cycle models that estimate based on the average speed of the journey to those that divide the journey into sections, and estimate the emissions in each section. The most detailed form of model is the instantaneous model that estimates over a very small time period (e.g. 1 second).

The National Atmospheric Emissions Inventory(NAEI) [2] is a United Kingdom based repository of tools for calculating the emissions from a wide range of activities, including a full-cycle vehicle emissions model. The model predicted emissions for the specified vehicle type (size, fuel type, engine class) for journey of a given duration.

COPERT [3,4] is a European Environmental Agency funded emissions model, which estimates vehicle emissions for a range of vehicles by engine classification (e.g. EURO1) and vehicle type such as passenger car, heavy goods vehicle. COPERT is driven by a database of emissions by vehicle class, engine technology and speed.

Both of the above models have drawbacks in that they do not model the journey with any accuracy. They assume a set of average speeds for all or large parts of the journey. A number of other emissions models exist, these include the United States EPA MOBILE [5] model, the EMFAC California [6] model and the German Handbook of Emissions Factors (HBDFA) [7].

A comparison of outputs from COPERT II, the German Handbook of Emissions Factors and the Danish DTU models may be found in [8]. This study compares the output from these models over a range of speeds and over a range of vehicles and shows that the predicted emissions differ between models. A comparison carried out in [8] suggests that COPERT predicts significantly less emissions than the Danish model, which predicts CO_2 emissions as being between 50% and 75% higher in specific instances. The focus of the paper being the effect of petrol engined catalyst equipped cars.

None of the above models are directly comparable with the Akcelik model being considered as they are not instantaneous models.

---

[1] Providing that training data can be obtained.

2.2 An Instantaneous Fuel Consumption Model

Akcelic and Biggs [9] proposed a fuel consumption and emissions model to be used in conjunction with micro simulation. By altering real-valued parameters used within the model the characteristics may be altered. These features give excellent potential to calibrate the model by making small adjustments to the parameters.

The model may be described as follows:

$$dF = \alpha dt + \beta_1 R_1 dx + [\beta_2 a R_1]_{a>0} \ for R_T > 0 \tag{1}$$
$$= \alpha dt \qquad\qquad for R_T \leq 0 \tag{2}$$

where dF = fuel (mL) consumed over distance dx (metres) during time dt(s), α = idle fuel rate (mL/s), β_1 = fuel consumption per unit of energy, β_2 = fuel consumption during positive acceleration, a = acceleration (m/s), negative when slowing down, and R_t = total force required to drive the vehicle (kN) expressed as follows:

$$
\begin{aligned}
R_t &= R_D + R_l + R_G,\\
R_D &= b_1 + b_2 v^2,\\
R_l &= Ma/1000,\\
R_G &= 9.81 M(G/100)/1000,
\end{aligned}
\tag{3}
$$

where: v is the speed (dx/dt) m/s, G is the gradient (%) +ve or -ve, M is vehicle mass (kg) and b_1, b_2 are the drag force function parameters. The total force required to drive the vehicle is $R_{(}t)$, calculated from drag $(R_{(}D))$, inertia $(R_{(}l))$ and grade$(R_{(}G))$

The model is used to indicate the fuel consumption of a period of time (typically 1 second) where the acceleration rate and gradient encountered, during the time period, are supplied. To convert fuel to CO_2 emissions, a conversion factor of 2.5g/ml (as specified in [10]) is used. Updated values for $\alpha,\beta_1,\beta_2,b_1$ and b_2 are provided in [10] to allow the model to be calibrated with respect to a Ford saloon car. The principle drawback to this approach however is the need to supply the parameters as described above, values of which, may not be readily available for different vehicles. In this study, we attempt to evolve parameters for the model based on data collected by an instrumented vehicle, a petrol engined Ford Mondeo estate car, driven over test routes within the city of Leeds, UK.

3 Data Sets

Data has been made available from the Institute for Transport Studies, Leeds University gathered as part of EPSRC Project (GR/S31136/01 RETEMM: Real-world Traffic Emissions Measurement and Modelling). The data consists of telematics and emissions data gathered from an instrumented vehicle. The vehicle was driven on a series of trips between three points. Speed and tailpipe emissions are gathered at intervals of 1 second. It is possible to retrospectively derive gradient, distance covered and speed information from the GPS data recorded.

The datasets used in this study represent a series of journeys made between three test sites. The first dataset $ES1$ contains 3526 one second readings, gathered over 10 journeys. Figure 1 shows the emissions readings contained in $ES1$ plotted against acceleration (based in the change in speed over each 1 second observation interval) . Observe the clear difference between emission characteristics between acceleration and deceleration. When the vehicle is decelerating the emissions rarely exceed 2gms per second). On the other hand, in acceleration, the emissions do not appear to be strongly related to the rate of acceleration. Note that in many cases the same rate of acceleration can produce different levels of CO_2 emissions. This clearly presents a challenge for any emissions model.

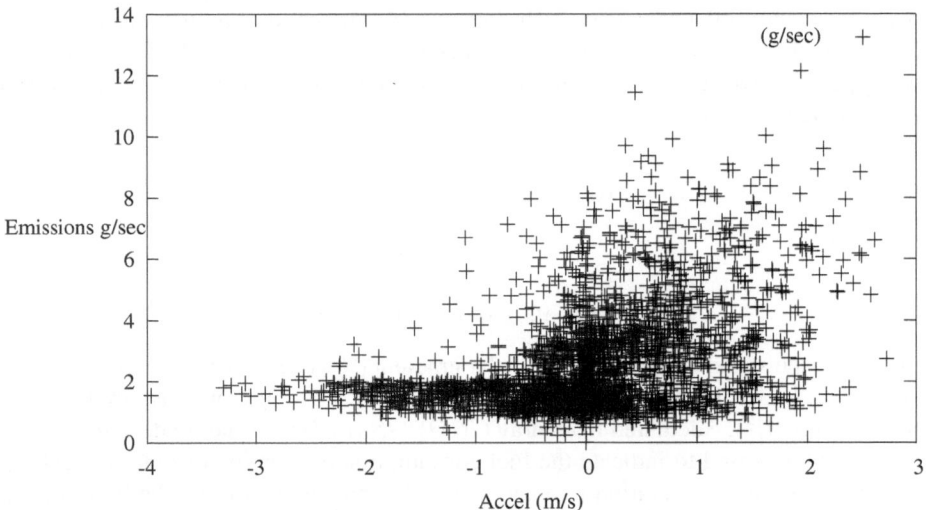

Fig. 1. Observed CO_2 tailpipe emissions and acceleration from the $ES1$ dataset (based on 1 second observations)

The algorithm was tested on a second dataset, based on a set of journeys made between the same three points. This dataset was divided into 10 specific journeys, known as $T1$ to $T10$. Each file represents one journey between two of the three test sites. The vehicle is stopped at each test site, so each file represents a journey from start to stop (most journeys will also include one or more intermediate stops due to traffic conditions). This division allows the reader to assess the performance of the algorithm on individual journey cycles.

A summary of the data used may be found in table 1.

3.1   Benchmarking the Model

Before addressing the use of an evolutionary algorithm to calibrate the model, it is useful to evaluate the model using the parameters specified in Akcelic 2003 ([10]) on the

Table 1. A summary of the datasets used in this paper

Name	Purpose	Qty of 1 second readings
ES1	Training set	3526
T1	Sample journey	951
T2	Sample journey	275
T3	Sample journey	310
T4	Sample journey	280
T5	Sample journey	280
T6	Sample journey	290
T7	Sample journey	290
T8	Sample journey	320
T9	Sample journey	330
T10	Sample journey	300

proposed test data sets. In [10], the model parameters are based upon those specified in [9] which are baed upon a Ford car with a non-catalyst 4.1l petrol engine and automatic transmission. This is a significantly different and older type of vehicle than the one used in the Leeds study (A 2005 Ford Mondeo, with a 1.7l Euro IV engine) and therefore it is expected that some difference in emissions would be observed.

However, the large % difference between the observed CO_2 and that estimated by the model given in table 2 of between 600% and 1800% suggests that there is massive scope for improvement in calibration of the model.

Table 2. Results obtained (% diff between actual and estimated) on the 10 datasets based on the original [10] parameters

Test set	T1	T2	T3	T4	T5	T6	T7	T8	T9	T10
% Difference	1814.89	618.62	618.64	576.91	703.4	605.8	609.78	743.99	692.54	677.62

4   The Evolutionary Algorithm

The purpose of the EA is to calibrate the model described in section 2.2 such that the model produced emissions estimates based on the vehicle characteristics contained in the observed data. This is achieved by determining appropriate values for the parameters α, β_1, β_2, b_1 b_2 (as defined in section 2.2). Note that M (mass) does not need to be determined as appropriate values are easily available.

A direct representation is employed that represents each of the five above parameters as double precision numbers . The parameters used within the algorithm were set arbitrarily as part of a tuning and testing process. The EA utilises a steady-state population of 200 individuals.The population was initialised randomly, no improvement was found through initialising some of the population using the parameters from [10], but some improvement was found through initialising a small number of individuals (10) with all values set to 0. In the results it was noted that some parameters take values close to 0., the author supposes that the inclusion of 0s aids the search for solutions with parameter values of close to 0.

Within each cycle a new individual is created, using crossover or by cloning a single parent, seleced using a tournament of size 2, with a probability of 0.7 or 0.3 respectively. Crossover creates a new individual using uniform crossover. A mutation is carried by randomly selecting one of the five genes and altering its value by adding, or subtracting, a random number drawn from a normal distribution N(0,x) where x is 10% of current value. The child individual is added to the population, replacing the loser of a tournament of size 2.

The fitness of an individual is determined by calculating the difference between actual and estimated emissions at each 1 second interval, the differences are squared and summed.

4.1 Results

Table 3 shows the results obtained from 10 separate runs of the algorithm. Note that 40% of the runs underestimated and 60% over estimated. Within the 10 sets of parameters evolved, little difference is noticed in their performance the variation in output being less than 1%. An output sample, showing the actual and estimated CO_2 output for part of journey $T1$ using the best parameters obtained is shown in figure 2. Note the apparent lag between the actual and estimated emissions, this may in some way be due to the physical characteristics of the vehicle, there may be a lag between tail pipe emissions altering and the acceleration profile of the vehicle. Further investigation is required in this area.

Table 3. Results obtained (% diff between actual and estimated, +VE figures represent over estimation -VE represent under estimation) when applying the best set of parameters obtained from 10 runs on the training data to each of the unseen datasets (T1 to T10)

	T1	T2	T3	T4	T5	T6	T7	T8	T9	T10
1	9.65	-8.21	4.94	-1.46	-12.08	2.43	-0.11	-12.12	0.34	-7.01
2	9.82	-8.05	5.08	-1.30	-11.94	2.57	0.03	-11.93	0.49	-6.85
3	9.76	-8.10	5.03	-1.36	-11.98	2.52	-0.01	-11.99	0.45	-6.90
4	9.71	-8.14	5.00	-1.40	-12.01	2.49	-0.05	-12.05	0.40	-6.95
5	9.83	-8.02	5.10	-1.28	-11.91	2.58	0.06	-11.91	0.52	-6.83
6	9.71	-8.14	4.99	-1.40	-12.02	2.48	-0.05	-12.04	0.40	-6.95
7	9.78	-8.10	5.03	-1.36	-12.01	2.48	-0.02	-11.99	0.46	-6.90
8	9.65	-8.19	4.96	-1.44	-12.05	2.46	-0.10	-12.10	0.35	-7.00
9	9.71	-8.15	4.99	-1.41	-12.04	2.46	-0.06	-12.05	0.40	-6.95
10	9.84	-8.02	5.10	-1.28	-11.91	2.58	0.06	-11.90	0.52	-6.82
Avg	**9.75**	**-8.11**	**5.02**	**-1.37**	**-11.99**	**2.51**	**-0.03**	**-12.01**	**0.43**	**-6.92**

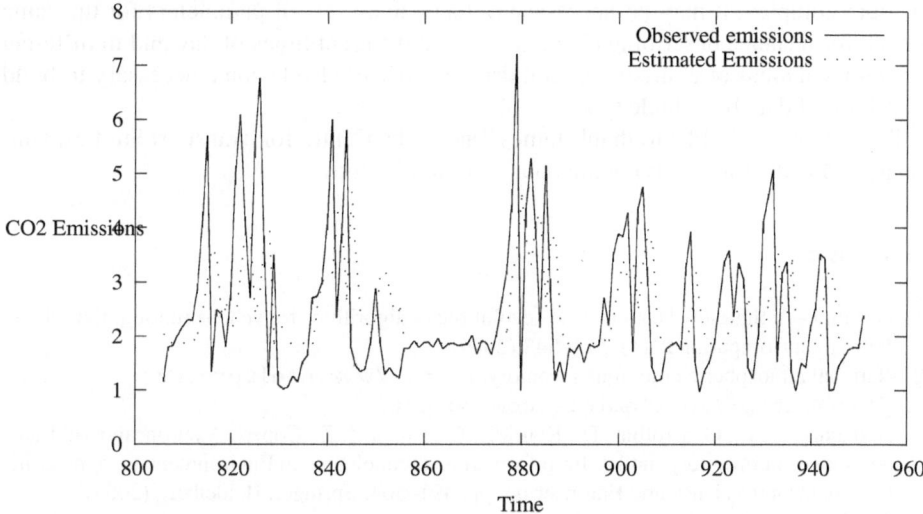

Fig. 2. An output sample, showing the actual and estimated CO_2 output for part of journey T1. The model parameters used were those shown in 3

5   Conclusions and Future Work

This paper provides a description of the work currently being undertaken which suggests potential for nature inspired methods to optimise emissions models by learning from observed data. It must be noted that at present both the training and test datasets are limited in scope. More experimentation is required with data that covers a wider range of road types and traffic conditions.

Although an Evolutionary Algorithm is utilised in this study, the author acknowledges that other approaches to solving this problem exist. Future work in this area should include the investigation of methods such as Grid Search [11].

It may be possible to determine values for α by examining emissions at the start of the journey whilst the vehicle is at rest and determining idle fuel consumption thus. This study covers petrol vehicles, as the emissions data available only covers a petrol engined motor car. It may be possible to adapt the model for other fuel/vehicle types (e.g. diesel engined truck) by using emissions data to re-calibrate the model and possibly alter the fuel to CO_2 conversion factor if another value is available.

The instantaneous model is of prime interest to those engaged in micro simulation of traffic. The next stage in this work is to investigate the possibilities of calibrating the instantaneous model to allow it to be applied to periods greater than 1 second. It may be possible to calibrate the model that it may be applied to longer, more complex sections of a journey. This may eventually allow the model to be used to estimate emissions in conjunction with Geographical Information Systems(GIS) systems, such as Google Maps which describe journeys as a series of variable length stages. The next stage of this work is to attempt to tune the parameters to work with such data to allow the evolved model to be used in conjunction with such systems. The problem may become far more complex, it may be necessary to have many sets of parameters for the same vehicle for instance describing characteristics at different times of day and in different geographical areas (e.g. urban or rural driving). It will also become necessary to build a database of data by vehicle type.

The author would like to thank James Tate at the Centre for Transport Studies, University of Leeds, for supplying the data used in this study.

References

1. Potvin, J.-Y.: State-of-the art review - evolutionary algorithms for vehicle routing. INFORMS Journal on Computing 21(4), 518–548 (2009)
2. National atmospheric emissions inventory, http://www.naei.org.uk/
3. Copert 4, http://www.emisia.com/copert/
4. Ntziachristos, L., Gkatzoflias, D., Kouridis, C., Samaras, Z.: Copert: A european road transport emission inventory model. In: Information Technologies in Environmental Engineering, Environmental Science and Engineering, pp. 491–504. Springer, Heidelberg (2009)
5. Mobile 6 (vehicle emission modeling software), http://www.epa.gov/oms/m6.html
6. California emissions factors (emfac), http://www.arb.ca.gov/msei/onroad/latestversion.html
7. The german handbook of emissions factors, http://www.hbefa.net/e/index.html
8. Winther, M.: Petrol passenger car emissions calculated with different emission models. The Science of The Total Environment 224(1-3), 149–160 (1998)
9. Akcelik, R., Biggs, D.C.: A discussion on the paper on fuel consumption modeling by post et al. Transportation Research Part B: Methodological 19(6), 529–533 (1985)
10. Akcelik, R., Besley, M.: Operating cost, fuel consumption, and emission models in aasidra and aamotion. In: 25th Conference of Australian Institutes of Transport Research (CAITR 2003). University of South Australia, Adelaide (2003)
11. Amos, N.A., et al.: The random grid search: A simple way to find optimal cuts. In: Prepared for International Conference on Computing in High-energy Physics (CHEP 1995), Rio de Janeiro, Brazil, September 18-22 (1995)

A Comparative Evaluation of Business Intelligence Technologies with Application to Product Profiling

Takudzwa Mabande, Joseph K. Balikuddembe, Antoine Bagula,
and Pheeha Machaka

Department of Computer Science
University of Cape Town
South Africa
{tmabande,jbalikud,bagula}@cs.uct.ac.za,
pheeha.machaka@uct.ac.za

Abstract. Most of the Business Intelligence tools available on the market today have either been developed and industrially operationalised as "one size fits all" solutions or offered with multiple options leaving the business to decide on the best technology to use. We infer that this approach is likely to result in various analysis inaccuracies; hence rendering inappropriate business decisions. Accordingly, evaluating which technologies present more accurate results against a particular business need remains imperative. While using customer data from a large financial services company in South Africa, we analysed the performance of Neural Networks, Artificial Immune Systems and Bayesian Networks in classifying customer buying patterns. We measured the accuracy percentage values for a customer's propensity to buy policies and also for existing policies lapsing. We observed that such assessments provide great insight in assessing the effectiveness of Business Intelligence enabling technologies. In particular, when applied to a larger data set, various customer patterns can be unearthed which results in adequate customer segmentation and business lead optimisation.

Keywords: Business Intelligence, Customer Profiling, Analysis Techniques.

1   Introduction

Business Intelligence (BI) is increasingly becoming popular. This popularity is driven by the need for allowing organizations to predict various business parameters such as markets, customer behaviour and sales; all with a degree of certainty [1]. Annually, customer loss results in significant loss of revenue to companies. Available literature shows that a small improvement in customer retention can lead to long-term profitability and success; in the least, it significantly improves profitability and market share [2].

Arguably, adopting technologies that can further such business strategies is rather embracive. However, these technologies must be well assessed to ensure that they are well tailored for the various needs. Sadly, the available solutions seem to assume a "one size fits all" mechanism. This has created various implementation disparities.

E. Hart et al. (Eds.): BIONETICS 2011, LNICST 103, pp. 133–140, 2012.
© Institute for Computer Sciences, Social Informatics and Telecommunications Engineering 2012

Prominently, evolving business models, unique data models, non standard reporting periods, non standard market definitions, dense aggregation models, high-level of interactive and ad-hoc reporting requirements, and dynamic event driven report scheduling requirements have validated this imbalance. On the application side, the perceived lack of 'what if' capabilities in BI tools, database performance issues and the business data not being integrated or aggregated enough still remain a critical concern[3].

Our work aims to assess which BI underlying technologies provide relatively accurate results with relation to a specific business need, customer profiling. Our hope is that this assessment will help organisations avoid spending money on tools that cannot provide an accurate and comprehensive analysis. To the best of our knowledge, the comparison we are making has never been made in the context of looking at what technology should fit what need. The rest of this paper is segmented as follows: section one comprises this introduction, section two looks at related work, section three describes our methodology, section four describes our results and analysis and section 5 describes our evaluation. We conclude with section 6 in which we provide future direction for this ongoing investigation.

2   Related Work

Literature shows that there are two perspectives to business intelligence understanding, that is, the managerial approach and the technical approach [4]. We focus on the technical approach which presents BI as a set of tools that gather, analyse and distribute information. Following are some examples of research done with these tools.

Lee and Shih [5] developed a neural network to recognise profitable customers for dental services marketing as a mechanism to promote dental implants in Taiwan. In their work, a multi-layer feed forward neural network, with a sigmoid function trained by back-propagation, was used. This network was able to achieve 90.18% classification accuracy for the test data provided. Baesens *et al* [6] propose predictor models for measuring customer's future spending patterns. Their results showed that an unrestricted Bayesian network present far better performance when measured against other methods used in validating their results. It had 75% classification accuracy in predicting the customer lifecycle. In another study, Baesens *et al* [7] focus on sales lead validation modelling. In this work, simulation results showed that the Bayesian neural network was suitable for repeat purchase modelling. It presented a correct classification result of 71% when compared against other techniques used such as logistic regression, linear and quadratic discriminant analysis.

We have not come across literature describing the application of Artificial Immune Systems in BI tools/solutions. However, this technique has widely been applied in classification and pattern recognition. For example, Harmer *et al* [8] and Lau *et al* [9] have applied it in intrusion detection. It has also been applied in collaborative Filtering and clustering [10 -11], more colloquially referred to as recommender systems.

Hence, this comparative in light of customer profiling and modeling; in conjunction with other widely applied techniques, can provide added benefit in analysis.

3   Our Methodology

3.1   Comparative Techniques Applied

3.1.1   Neural Networks (NN)
Cross *et al* [12], define a NN as a set of simple computing elements called nodes that are interconnected via a set of weights. These weights are adjusted when data is presented into the network during a training process to enable classification. There are different ways of training a NN, for example back-propagation. However, due to drawbacks of gradient descent training mechanisms, we used a genetic algorithm in training our NN. Previous work, for example [13], shows that Genetic algorithms can outperform back-propagation considerably.

3.1.2   Bayesian Belief Networks (BN)
Jansen and Jansen [14] defined BNs as a set of variables each with a finite set of mutually exclusive states forming a directed acyclic graph where each variable with parents has an associated probability table. BNs have been proven to be powerful tools in knowledge representation and inference under uncertain conditions but can suffer from scalability issues [15-16]. In this work, we used a constraint based learning algorithm to learn the structure of the Bayesian Network. This was implemented using the Recursive Autonomy Identification method proposed by Yehezkel and Lerner [17]. This was specifically chosen because of its increased stability and decreased computational complexity.

3.1.3   Artificial Immune Systems (AIS)
Harmer *et al* [9] define AISs as computational systems that are inspired by theoretical immunology and observed immune functions, principles and models; which are applied to complex problem domains. They are modelled or derived from Natural Immune Theory. The choice of which AIS technique to use was based on the various aspects of immunology and on the type of data and queries in question. These techniques include Negative (and Positive) Selection, Danger Models, Clonal Selection and Immune Network Models. In this work we applied the Clonal AIS model proposed by Garret [10].

3.2   Case Study Description

We obtained data from one of the largest financial service firms in South Africa. This company was looking at growing its BI competencies to boost its market share. It was looking at analysing historical and current view presentations of its business operations, specifically within the area of sales lead validation. Its current BI approach lacked the predictive or forecasting views of analysing these presentations. Thus, developing accurate forecasts became a key requirement for this project.

3.3   Architectural Design

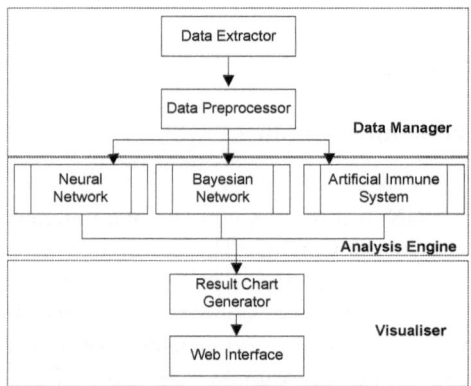

Fig. 1. Architectural System components

3.3.1   Experiment

3.3.1.1   Characteristics of Input Parameters. Data provided by our industrial partner contained limited customer personal information and purchasing history. This data formed the Customer Object in this analysis.

3.3.1.2   Data Pre-Processing. The dataset contained instances with missing attributes which were excluded from classification. Nominal attributes were binarised and the dates were converted to the duration of holding the policy. Every attribute was normalised to a value between 0 and 1.

3.3.1.3   Test Execution Steps. All the three techniques were tested against each other. The results obtained were then validated against techniques available in the Weka open source package [18]. Performance was evaluated based on the following questions.

1.   Which technique achieves better accuracy given this kind of data?
2.   How does each technique compare in terms of predicting whether customers will buy a certain policy?
3.   How does each technique compare in terms of predicting which policies are likely to lapse?

Test 1: We used both a simple split and 10 –fold-cross-validation. For the split test 66% of the data was used as a training set and the remaining 34% as the test set. We applied all the techniques in predicting whether a given policy is currently active or inactive.

Test 2: A customer with a certain set of attributes was given as input and the customers` propensity to buy certain policies was returned. Only propensities above a certain threshold were returned in the output. We then compared and analysed the results thereafter.

Test 3: This test was similar to test 2. In this case though we were looking for which customer was likely to lapse on their current policy.

4   Results

We obtained the following accuracy results as given in table 1 below.

Table 1. accuracy percentages of all machine learning techniques

Technique	Percentage Accuracy
GA Neural Network	80.40%
Artificial Immune System	76.20%
Weka Back Propagation Neural Network	79.43%
Weka Multilayer Perceptron	81.97%
Weka Bayesian Network	80.71%
Weka Artificial Immune System	70.11%

As illustrated in table 1 above, there was no large difference in accuracy between techniques. The Weka implemented AIS scored lowest; but all the other techniques had over 76% accuracy. When looking at the confusion matrices, these accuracy percentages could partly be attributed to the nature of the data.

For instance, the GA NN and the Bold Driver Back Propagation algorithm both classified almost all instances of policies as lapsed policies. With the exception of the Weka BN, all the other machine learning techniques had a bias towards inactive classification. Reasonable accuracies were still obtained as 79.43% of the data was composed of lapsed policies. This imbalance and a lack of easily separable clusters meant a large bias towards classifying all policies as lapsed to optimize the error function. The rest of the results will be based on the most accurate implementation of each technique.

The NN had a kappa statistic of 0.3088 compared to 0.3197 for the BN and 0.1418 for the AIS. The kappa statistic indicates the probability of a technique being able to identify a renewed policy taking into account random chance. It is an indication of whether the technique is any better than random chance. This statistic shows that the BN performed better than the other two techniques and the AIS was especially poor.

Using three other measures, precision, recall and the f-measure, the performance of each technique can be measured against each possible outcome (lapsed or renewed). Precision is how accurate the technique is at identifying that a certain customer is likely to lapse or renew a policy. Recall is the ratio of customers it correctly classifies as lapsing or renewing to the total customers lapsing or renewing. Recall is a measure of relevance whilst precision is a measure of accuracy. A precise technique will classify only customers lapsing on their policies as lapsing, but it may have poor recall in that not all customers lapsing will be classified as lapsing (e.g. all 11 customers classified as lapsing are truly lapsing on their policy, however there are 30 customers in the data set lapsing on their policies). F-measure is the harmonic mean of precision and recall. It is used to assess a techniques performance in classification for a given class.

The NN had a precision of 0.839 for lapsing policies but 0.636 for renewed policies. The BN was slightly more accurate at identifying lapsed policies with a precision of 0.846 but less accurate with renewed policies with a precision of 0.549. The AIS was almost as precise with 0.827 for the lapsed policies but was much poorer at identifying renewed policies with a precision of 0.308.

The NN had a recall of 0.957 for the lapsed policies and 0.289 for the renewed policies. The BN had a slightly lower recall of 0.925 for the lapsed policies and 0.352 for the renewed policies. The AIS had the lowest lapsed recall of 0.789 but 0.362 for renewed policies. The significant difference in lapsed policy recall explains why the AIS had a much lower overall accuracy even though its precision was only slightly lower than the other two techniques. Even though the AIS had the best renewed policy recall, the lapsed policies formed a much larger proportion of the data set and therefore impact heavily on overall accuracy.

The NN had an f-measure of 0.894 for the lapsed policies and 0.398 for the renewed policies. The BN had f-measures of 0.884 and 0.429 for lapsed policies and renewed policies respectively. The AIS had the lowest f-measures of 0.807 for the lapsed policies and 0.333 for the renewed policies. This shows that the BN was the best at identifying renewed policies but slightly less than NN at identifying lapsed policies. The NN had a better overall accuracy because of the larger ratio of lapsed policies in the data set mentioned above. The AIS did not perform as well as the other two techniques.

Additionally, an analysis of randomly selected propensities per customer provided by the NN and the AIS yielded the following data illustrated in table 2 below.

Table 2. An analysis of the propensities produced by the profiling techniques

Groups	Count	Sum	Average	Variance
AIS	158	95.01	0.601329	0.02425
NN	158	78.4102	0.496267	0.000396

The table shows that the AIS had a wider range of propensities, with a significantly higher average propensity. Thus, it was more optimistic; identifying some customers as likely to buy with a propensity/probability of one.

An ANOVA test was performed on the propensity data using a 0.0001 confidence level. The p-value of $1.44e^{-5}$ was attained. This was measured as the probability that the calculated value might be observed against the given propensities of the NN and the AIS; with the assumption that all are the same. This showed that there is a similarity of the propensities of the two techniques, however it is quite small.

5   Evaluation

These results showed that no significant difference existed in overall accuracy between the techniques applied. Additionally all the techniques could not find a clear distinction between a customer who renewed policy, and a customer who lapsed on their policy. Firstly, the NN was biased towards lapsing all policies; probably due to the optimization of the respective error functions. The K2 search BN was much better at identifying renewed policies, but was prone to type one errors (false positives). The constraint based Bayesian Network implementation failed to train on the customer data. The AIS technique was overly optimistic, mostly giving high policy buying propensities. This meant a significant number of type one errors in the analysis. The AIS was not as good as the other techniques in almost all metrics.

More still, the propensities showed that there was some correlation in the results attained by the AIS and the NN. The propensities were higher for the AIS. The low values for the NN are accounted for by the bias towards lapsing. The NN was much faster in the batch tests than the AIS because NNs can be trained once on a given dataset. The weights generated are useful until there are significant changes to the data. Although AISs require less resources in training, the need for them to train on every run means that the NN in this case outperformed the AIS in terms of speed on a query.

6   Conclusions and Future Work

As the demand for BI enabling technologies capable of supporting sustainable business growth rises, the need for development of BI tools that are less resource intensive, providing accurate predictions and also performing with in reach will be unprecedented. End-users are more interested in BI applications which are less complex in terms of use, having a significant degree of data democracy and relatively fast in terms of performance. The analysis presented in this work aimed at supporting this view.

Although we stumbled on a number of technique bottlenecks in our investigation, there are issues observed that will be interesting to investigate in future. Domain knowledge integration in supporting the establishment of BI initiatives opens up a new interdisciplinary research area between business and technology. Equally, automated attribute selection techniques could be used to improve performance. Further comparison of other BI enabling techniques, including hybrids of those used in this research could perhaps provide more insight in this area, especially if more datasets are used. Hence, our hope is that our results can help implementers to align different analysis needs with the appropriate technologies at hand.

References

1. Jourdan, Z., Rainer, R.K., Marshall, T.E.: Business Intelligence: An Analysis of the Literature. Inf. Syst. Manage. 25, 121–131 (2008)
2. Buckinx, W., Van den Poel, D.: Customer base analysis: partial defection of behaviourally loyal clients in a non-contractual FMCG retail setting. Eur. J. Oper. Res. 164, 252–268 (2005)

3. Chung, W.: Enhancing Business Intelligence Quality with Visualization: An Experiment on Stakeholder Network Analysis. Pacific Asia Journal of the Association for Information Systems 1(1), Article 9 (2009)
4. Petrini, M., Pozzebon, M.: Managing sustainability with the support of business intelligence: Integrating socio-environmental indicators and organisational context. J. Strat. Inf. Syst. 18, 178–191 (2009)
5. Lee, W.I., Shih, B.Y.: Application of neural networks to recognize profitable customers for dental services marketing-a case of dental clinics in Taiwan. Expert Syst. Appl. 36, 199–208 (2009)
6. Baesens, B., Verstraeten, G., Van den Poel, D., Egmont-Petersen, M., Van Kenhove, P., Vanthienen, J.: Bayesian network classifiers for identifying the slope of the customer lifecycle of long-life customers. Eur. J. Oper. Res. 156, 508–523 (2004)
7. Baesens, B., Viaene, S., Van den Poel, D., Vanthienen, J., Dedene, G.: Bayesian neural network learning for repeat purchase modelling in direct marketing. Eur. J. Oper. Res. 138, 191–211 (2002)
8. Kim, J., Bentley, P.J.: Towards an artificial immune system for network intrusion detection: an investigation of clonal selection with a negative selection operator. In: Proceedings of the 2001 Congress on Evolutionary Computation, vol. 2, pp. 1244–1252 (2001), doi:10.1109/CEC.2001.934333
9. Harmer, P.K., Williams, P.D., Gunsch, G.H., Lamont, G.B.: An artificial immune system architecture for computer security applications. IEEE Transactions on Evolutionary Computation 6(3), 252–280 (2002), doi:10.1109/TEVC.2002.1011540S
10. Garrett, S.M.: How Do We Evaluate Artificial Immune Systems? Evol. Comput. 13(2), 145–177 (2005)
11. Glickman, M., Balthrop, J., Forrest, S.: A Machine Learning Evaluation of an Artificial Immune System. Evol. Comput. 13(2), 179–212 (2005)
12. Cross, S., Harrison, R.F., Kennedy, R.L.: Introduction to neural networks. The Lancet 346, 1075–1079 (1995)
13. Sexton, R.S., Dorsey, R.E.: Reliable classification using neural networks: a genetic algorithm and backpropagation comparison. Decis. Support Syst. 30, 11–22 (2000)
14. Neapolitan, R.E.: Learning Bayesian Networks. Pearson Prentice Hall (2004)
15. Cheng, J., Greiner, R.: Comparing bayesian network classifiers. In: Proceedings UAI (1999)
16. Kohavi, R.: Scaling up the accuracy of naive-bayes classifiers: A decision-tree hybrid. In: Proceedings of the Second International Conference on Knowledge Discovery and Data Mining (1996)
17. Yehezkel, R., Lerner, B.: Bayesian Network Structure Learning by Recursive Autonomy Identification. Journal of Machine Learning Research 10 (2009)
18. Hall, M., Frank, E., Holmes, G., Pfahringer, B., Reutemann, P., Witten, I.H.: The WEKA Data Mining Software: An Update. SIGKDD Explorations 11(1) (2009)

A Biologically-Inspired Model for Recognition of Overlapped Patterns

Mohammad Saifullah

Department of Computer and Information Science
Linkoping University, Sweden
Mohammad.saifullah@liu.se

Abstract. In this paper a biologically-inspired model for recognition of overlapped patterns is proposed. Information processing along the two visual information processing pathways, i.e., the dorsal and the ventral pathway, is modeled as a solution to the problem. We hypothesize that dorsal pathway, in addition to encoding the spatial information, learns the shape representations of the patterns. In our model dorsal pathway uses shape knowledge as a top-down guidance signal to segment the bottom-up saliency map of the overlapped patterns. Segmented map is used to modulate processing in the ventral pathway. Pattern segmentation in the dorsal pathway is implemented as an interactive process. Interaction between the two pathways leads to sequential recognition of the overlapped patterns along the ventral pathway, one after another. Simulation results support the presented hypothesis as well as effectiveness of the model.

Keywords: Vision, Attention, Neural network model, Overlapped patterns, Biologically-inspired, Saliency map, Interactive process.

1   1 Introduction

Recognition of overlapped pattern is a complex computational problem. This problem involves two fundamental issues in the field of machine vision, i.e., pattern segmentation and pattern recognition. The main difficulty in the recognition of overlapped patterns is accurate segmentation of individual patterns.

Traditional visual information processing theories [1][2], consider the segmentation and recognition as two independent process and, suggest that segmentation proceeds and facilitates recognition. In the same way, most of the computational vision models [2][3] tackle the problem of overlapped patterns with a bottom-up, 'Recognition followed by segmentation', approach. In this approach segmentation is performed by using image-based information only. But, these bottoms-up approaches cannot perform segmentation satisfactorily in the absence of top-down cues. There are evidences from the human vision studies [4] that high-level, class specific cues play an important role in image segmentation.

Human visual system has very good object recognition ability under large variations in conditions. In the event of many objects in the visual field, the phenomenon of visual

E. Hart et al. (Eds.): BIONETICS 2011, LNICST 103, pp. 141–148, 2012.
© Institute for Computer Sciences, Social Informatics and Telecommunications Engineering 2012

attention helps to select and process one object at a time. Visual attention can be best understood in terms of two stream hypothesis. These streams are the ventral stream or 'What' pathway and the dorsal stream or 'Where' pathway. The ventral stream is responsible for recognizing and identifying objects while the dorsal deals with the spatial information associated with objects. The interaction between these two pathway lead to focus of attention on one of the many objects present in the scene. In this work we hypothesize that the dorsal channel learns the shape knowledge of the patterns to segment saliency map of the overlapped patterns as an interactive process. Recent studies in neuroscience [5][6] indicates that neurons in lateral intraparietal cortex (LIP) part of the dorsal pathway show some selectivity for shapes of the patterns, and hence suggest the possibility of encoding shape information along the dorsal pathway.

Most of the neural network models have used the bottom-up approaches for image segmentation. Mozer and colleagues [7] proposed a neural network model to segment overlapped objects. The network learns the underlying grouping principal to segment the objects. This model is a basically a bottom-up approach as it does not utilize any object knowledge to guide segmentation. Grossberg [8] presented a biologically-inspired model of the visual cortex that approaches the figure-ground separation as an interactive process. Another biologically-based neural network model is MAGIC [9], developed by Behrmann and colleagues, which uses bidirectional connections to model grouping of features into objects within the ventral pathway. A closely related approach to our work is proposed by Vecera and O'Reilly [10][10]. In their work they demonstrated the figure-ground segmentation as an interactive process.

2   Approach

The approach used in this work is inspired by, generally, the information processing in the human brain and more specifically by the phenomenon of visual attention. Here, the strategy is to develop a saliency map in the dorsal channel that represent the spatial as well as identity information of one of the overlapped pattern at a time and then use this information to modulate processing along the ventral channel in favor of that pattern.

The main thrust of this approach is on incremental development of a saliency map as an interactive process: 'recognition followed by segmentation, segmentation followed by recognition'. In this approach a salient patch of the bottom-up saliency map of the occluded patterns is selected and identified as a part of a specific pattern. This higher level perception then guide the growth of saliency map in a proximate region (pixels) that belong to the same pattern. This step of incremental growth by using the bottom-up image information and top-down guidance of the pattern shape information proceeds in small steps and lead to a map that encode the spatial location as well as shape information of one of the occluded pattern.

3   The Model

It is a hierarchical multi-layered neural network as illustrated in the figure 1. The model has two parallel processing channels: the ventral and the dorsal channel.

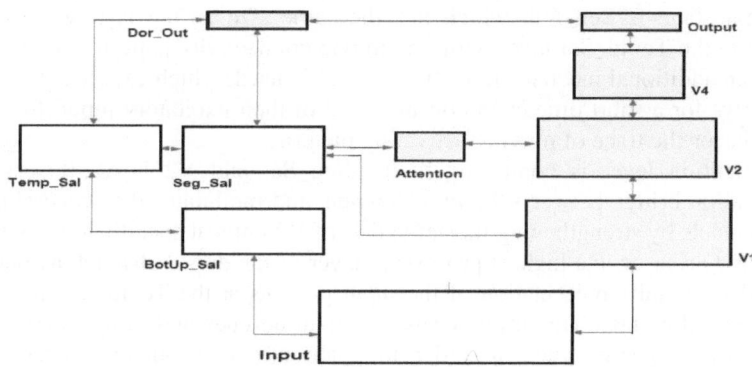

Fig. 1. The proposed biologically-inspired model

The Ventral Channel: This channel learns invariant representations of the input patterns while processing information along its hierarchical structure.

The V1 layer, vey first layer in this pathway, is designed to detect oriented edge like biologically-inspired features from the input image. The units in V1 layer are arranged into groups such that each group processes a specific part of the input image. The next layer along the hierarchy is V2, having larger receptive field size than that in the previous stage. The units of this layer are arranged into groups and extract complex features from the previous layer. The next layer in the hierarchy is V4, which has the largest receptive field in the hierarchy that encompasses the whole input at the Input layer. Every unit of this layer gets input from all units in the previous layer. The Output layer is the last and highest processing stage of this channel. It learns to assign the different representations of same class at V4 to a single distinct output unit.

The Dorsal Channel: This channel encodes spatial as well as shape properties of the input in the form of saliency map, and use this information to guide segmentation of overlapped patterns. This pathway interacts with the ventral pathway, at appropriate levels of processing stages, to modulate it in favor of specific locations and features.

The first stage of processing in this channel is BotUp_Sal layer, which develops bottom-up saliency map of the input. It is smaller in size than the Input layer and each unit of this layer receive input from a group of units from the Input layer. This stage processes information on the basis of its connectivity patterns and does not perform any learning.

The next stage of this channel, Temp_Sal layer, is of the same size as BotUp_Sal and connected through it with a one to one connectivity pattern. Moreover, each unit of this layer is connected to eight close neighbors through self-connection. This layer , due to high value of K for KWTA, allow only a very small number of units to be active at a given time, furthermore, the self-connectivity force the active units to appear in the form of one or two groups. Additionally, a mechanism of 'accommodation' was employed which causes a unit that is active for a while to get tired and die down for a short period of time thereby giving other competing units to become active.

The next stage is Seg_Sal, which has the same size as the last two layers and connected to the Temp_Sal layer with one to one connectivity pattern among units. In this layer an additional mechanism of 'hysteresis' is used, which causes active units to remain active for a short time even after removal of their excitatory input. In this way, this layer keeps the trace of moving activating patterns.

The Attention layer is connected to the Seg_Sal and V2 layer. It serves as a communication bridge between the two channels and modulates the processing in the ventral channels by strengthening the activation of V2 units at specific locations.

The Dor_Out layer, the highest processing layer of the dorsal channel, associates the template like spatial representation of the input patterns at the Temp_Sal and Seg_Sal layer to a specific class. This layer receives bottom-up cues and in response generates top-down shape specific cues, in order to guide the segmentation processes in an interactive manner. This layer is also connected to the Output layer of the ventral channel.

3.1   Network Algorithm

An interactive (bidirectional, recurrent) neural network was developed in Emergent [11] using the biological plausible algorithm Leabra [12]. Each unit of the network had a sigmoid-like activation function. Learning in the network was based on a combination of Conditional Principal Component Analysis (CPCA), which is a Hebbian learning algorithm and Contrastive Hebbian learning (CHL), which is a biologically-based alternative to back propagation of error, applicable to bidirectional networks [12].

$$\text{CPCA:}\quad \Delta_{\text{hebb.}} = \varepsilon y_j(x_i - w_{ij}) \tag{1}$$

ε = learning rate
x_i = activation of sending unit i
y_j = activation of receiving unit j
w_{ij} = weight from unit i to unit $j \in [0, 1]$

$$\text{CHL:}\quad \Delta_{\text{err}} = \varepsilon(x_i^+ y_j^+ - x_i^- y_j^-) \tag{2}$$

x^-, y^- = act when only input is clamped
x^+, y^+ = act when also output is clamped

$$\text{L\_mix:}\quad \Delta w_{ij} = \varepsilon[\, c_{\text{hebb}} \Delta_{hebb} + (1 - c_{\text{hebb}})\Delta_{err}] \tag{3}$$

c_{hebb} = proportion of Hebbian learningIn Eq 3, L_mix represent the weight update as a result of learning that is a combination of Hebbian and error-driven learning.

Fig. 2. Top row: Six English alphabet pattern classes used for simulations. Bottom row: Some of the test patters developed by overlapping the patterns.

4   Simulations and Results

To simplify training and analysis of the results gray level images of six English alphabet patterns (fig. 2, top row) were used for simulations. Each of the patterns was shifted two steps in all eight directions, where length of the each step is four pixels. For testing, stimuli containing overlapped patterns were created (fig. 2, bottom row). Such that, each stimulus contains two randomly selected patterns.

Training of the two network channels, i.e., the ventral and the dorsal channels was performed separately. In the following section a few simulation results with brief analysis are presented to demonstrate the workability of the approach.

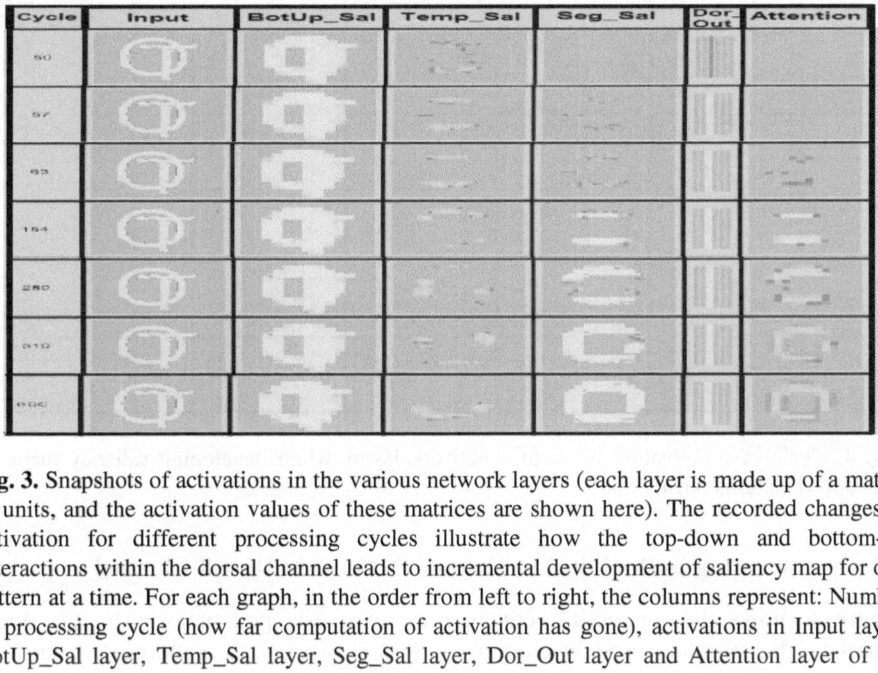

Fig. 3. Snapshots of activations in the various network layers (each layer is made up of a matrix of units, and the activation values of these matrices are shown here). The recorded changes in activation for different processing cycles illustrate how the top-down and bottom-up interactions within the dorsal channel leads to incremental development of saliency map for one pattern at a time. For each graph, in the order from left to right, the columns represent: Number of processing cycle (how far computation of activation has gone), activations in Input layer, BotUp_Sal layer, Temp_Sal layer, Seg_Sal layer, Dor_Out layer and Attention layer of the network. Yellow (light) colors denote high activation values, red (dark) colors low activation. Gray (neutral) color means no activation.

4.1   Segmentation of One of the Overlapped Pattern in Dorsal Channel

Figure 3 shows the behavior of the dorsal channel when a stimulus is presented to the network. This stimulus contains two overlapped patterns 'O' and 'T'. A saliency map of the overlapped pattern is generated at the BotUp_Sal layer (cycle: 50). Next, Temp_Sal layer get activation from the BotUp_Sal layer, and due to low value of K for the KWTA relatively small number of its units become active. Due to self-connection, the active units in this layer appear in the form of one or two groups (cycle: 57-63). The Temp_Sal layer, on the basis of the similarity of the active groups with the learned pattern templates, activates the Dor_Out layer that represents, in this instance, the pattern 'O'. The Dor_Out layer in return feedback to the Temp_Sal and Seg_Sal layers, and thus modulate these in favor of pattern 'O'. This interactive process proceeds in small steps and consequently active group of units at Temp_Sal layer moves over the spatial area representing pattern (obtaining input from saliency map of the 'O' in the BotUp_ Sal layer) 'O' (57-600). In the meantime, the Seg_Sal layer keeps the traces of the moving groups at the Temp_Sal layer (cycle: 67-600). The resulting map at the Seg_Sal layer appears in the form of a rough template representing the pattern 'O'.

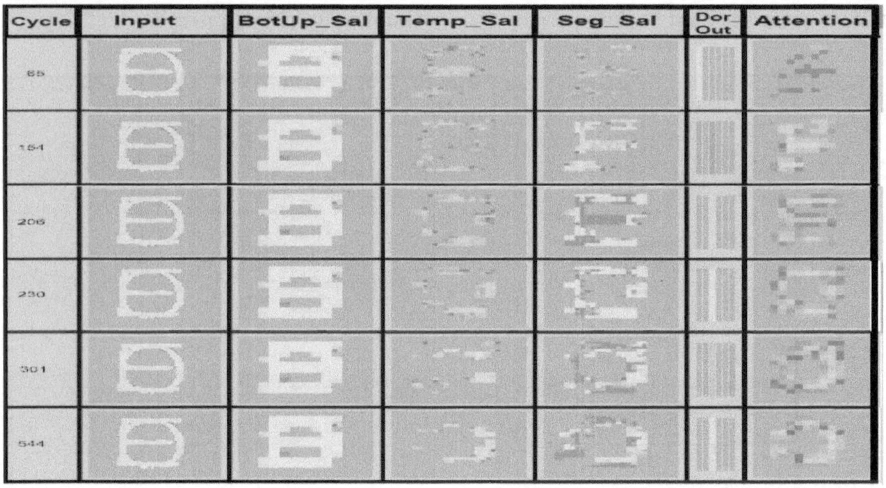

Fig. 4. Cycle-wise activation of various network layers while developing saliency maps for overlapped patterns one by one

4.2   Segmenting Both of the Overlapped Patterns in Dorsal Channel

Figure 4 shows the sequential segmentation of overlapped patterns by the dorsal channel. The stimulus used is composed of two overlapped patterns 'E' and 'O'. When it is presented to the model, the BU_Map generate a bottom-up saliency (cycle: 12-544). The Temp_Sal layer activates a chunk of this saliency map which subsequently identified as the pattern 'E' at the Dor_Out layer (cycle: 65-154). As a result, the top-down signals are generated that modulate the Temp_Sal and Seg_Sal layers in favor of pattern 'E'. As bottom-up signals contain the activation patterns of 'E', therefore, Seg_Sal grows and develops a map for 'E' due to interaction between top-down and bottom-up influences (cycle: 65-206). The attention layer also shows the corresponding saliency maps (cycle: 65-206). Meanwhile, some of the active units at the Seg_Sal layer start getting tired and thus give way to other competing units to become active (cycle: 154-206). Most probably the newly activated units belong to the other pattern 'O', because units that represents 'E' were already active. At this moment the newly active units support the pattern 'O' and thus forced the Dor_Out

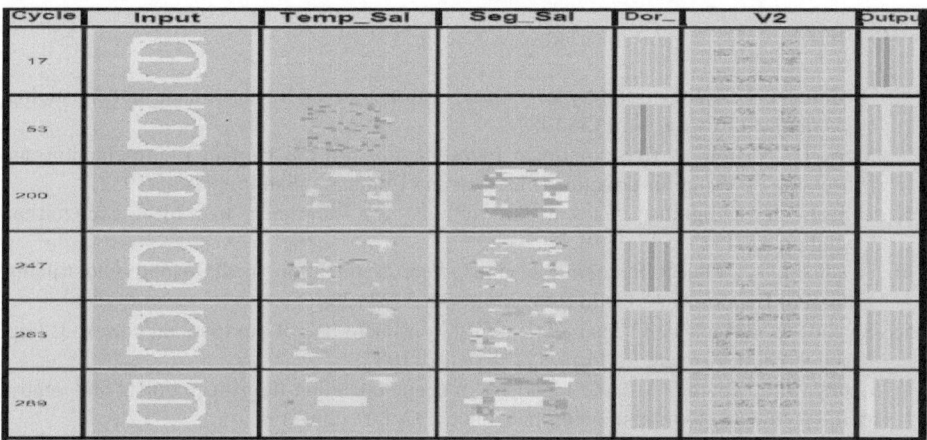

Fig. 5. Cycle-wise **activation of various network layers while developing saliency maps for** overlapped patterns one by one, and correspondingly modulation and recognition of patterns in the ventral channel

layer to generate top-down signals that favor this new pattern (cycle: 154). Consequently, the top-down and bottom-up interaction leads to development of a saliency map that represents the pattern 'O' (cycle: 230-544).

4.3   Segmentation in Dorsal and Recognition in the Ventral Channel

In this simulation interaction between the dorsal and the ventral channel leads to sequential recognition of overlapped patterns in the ventral channel (fig. 5). This simulation is an extension of the last simulation, i.e., same stimulant is used as in the last simulation, and dorsal channel perform the similar processing, but this time dorsal channel is connected to the ventral channel. Consequently, the dorsal channel and the ventral channel interact at three levels: Seg_Sal and Attention layers modulate V1 and V2 layers, respectively, for the specific spatial locations, and Dor_Out layer influence the Output layer of the ventral channel, which in turn provides feature based modulation along the ventral channel. This interactive processing leads to sequential recognition of overlapped patterns by the ventral channel as 'O' (cycle: 63-247) and later 'T' (cycle: 263-289).

5   Conclusion

In this paper a biologically inspired model for the recognition of overlapped patterns is presented. In the dorsal pathway, an interactive process based on top-down shape information and bottom-up image information, segments the bottom-up saliency map of the overlapped patterns. The interaction between the dorsal and the ventral pathway modulate the processing in favor of one pattern at a time. The computers simulations with the model demonstrate that it provides a satisfactory solution to the problem of overlapped patterns recognition which has little bottom-cues for segmentation.

References

[1] Biederman, I.: Recognition-by-components: A theory of human image understanding. Psychological Review 94, 115 (1987)
[2] Marr, D.: Vision. A Computational Investigation into the Human Representation and Processing of Visual Information. Freeman and Company, San Francisco (1982)
[3] Ullman, S.: Aligning pictorial descriptions: An approach to object recognition. Cognition 32, 193–254 (1989)
[4] Needham, A.: Object Recognition and Object Segregation in 4.5-Month-Old Infants. Journal of Experimental Child Psychology 78, 3–22 (2001)
[5] Sereno, A., Maunsell, J.: Shape selectivity in primate lateral intraparietal cortex. J. Exp. Psychol. Hum. Percept. Perform. 12, 388–391 (1987)
[6] Lehky, S.R., Sereno, A.B.: Comparison of shape encoding in primate dorsal and ventral visual pathways. Journal of Neurophysiology 97, 307 (2007)
[7] Mozer, M.C., Zemel, R.S., Behrmann, M., Williams, C.K.I.: Learning to Segment Images Using Dynamic Feature Binding. Neural Computation 4, 650–665 (1992)
[8] Grossberg, S.: A Solution of the Figure-Ground Problem for Biological Vision. Neural Networks 6, 463–483 (1993)
[9] Behrmann, M., Zemel, R.S., Mozer, M.C.: Object-based attention and occlusion: Evidence from normal participants and a computational model. Journal of Experimental Psychology: Human Perception and Performance 24, 1011 (1998)
[10] Vecera, S.P., O'Reilly, R.C.: Figure-ground Organization and Object Recognition Processes: An Interactive Account. Journal of Experimental Psychology: Human Perception and Performance 24, 441 (1998)
[11] Aisa, B., Mingus, B., O'Reilly, R.: The Emergent Neural Modeling System. Neural Networks 21, 1146–1152 (2008)
[12] O'Reilly, R.C., Munakata, Y.: Computational Explorations in Cognitive Neuroscience: Understanding the Mind by Simulating the Brain. The MIT Press (2000)

A Bio-inspired Coverage and Connectivity Maintenance Algorithm

Emi Mathews, Tobias Graf, and Kosala S.S.B. Kulathunga

University of Paderborn
Paderborn, Germany
emi@hni.upb.de, {tobiasg,suranga}@mail.upb.de
http://www.uni-paderborn.de/eps

Abstract. Swarm robots provide greater flexibility and robust performance in tasks such as sensing and monitoring of unstructured and unpredictable environments. They need to spread out in these environments maximizing *coverage* and maintaining network *connectivity* for efficient operation. Inspired from nature, we design a new coverage and connectivity maintenance algorithm. The algorithm is based on the local rules used by fish while schooling. Each robot is subject to three forces: a) A *separation force* that pushes it away from its neighbours and increases the size of the swarm. b) A *cohesion force* that maintains the connectivity of the swarm. c) An *alignment force* that keeps it aligned to its neighbours and makes relocation faster. Empirical analysis shows that our new algorithm improves coverage and maintains connectivity. Moreover, preliminary results obtained from the basic experiments show that the new swarm-based algorithm outperforms even the most prominent state-of-the-art algorithms, achieving better and faster coverage.

Keywords: swarm robots, bio-inspired, fish schooling, coverage, connectivity.

1 Introduction

Swarm robotics has gained significant attraction of robotic researchers in the domain of multi-robot coordination and control, due to their capability to perform complex behaviours at the macro-level, with high level of fault tolerance and scalability. In this paper, we focus on using swarm robots to spread out an area for sensing and monitoring applications. In areas such as disaster fields, toxic urban regions, subterranea or remote planets, human interventions are not possible or very dangerous. Using swarm robots for sensing and monitoring, provide greater flexibility and robust performance in such unstructured and unpredictable environments.

When the swarm robots spread out an area, an interesting problem to look at is the *coverage* maximization problem, where coverage is defined as the ratio of covered area of all robots to the total area to be covered [2]. Another important aspect to consider while spreading out is to keep the network of robots connected.

E. Hart et al. (Eds.): BIONETICS 2011, LNICST 103, pp. 149–154, 2012.

Thus the problem of *coverage* maximization maintaining the *connectivity*, denoted as $C - C$, arises when the robots spread.

The $C-C$ problem has been studied previously in the fields like wireless sensor networks and mobile robotics [3,4,6,9]. In this article, we focus on designing a bio-inspired algorithm for solving the $C - C$ problem. Bio-inspired algorithms work well in environments where prior-knowledge about the environment is minimal. They adapt to unforeseen changes in the task environment quickly. Hence, they are the most suitable choice for swarm robotic control and coordination.

The main idea used in our coverage and connectivity maintenance algorithm is born from the schooling behaviour of fish which "optimize" the $C-C$ constraint naturally, i.e.

- The swarm needs to stay together to appear as one fish. So they maintain connectivity.
- The appearance needs to be as large as possible to frighten predators. So they try to increase the coverage.

Thus, nature offers a natural concept for solving the $C - C$ problem. We look at this natural model and design our new coverage and connectivity maintenance algorithm. We show in this paper that this bio-inspired approach for swarm robotic control and coordination achieves better and faster coverage.

The remainder of this paper is organized as follows: Section 2 presents our new coverage and connectivity maintenance algorithm. In Section 3, the empirical analysis to evaluate the performance of the proposed algorithm and the preliminary results are discussed. Finally, Section 4 summarizes the work and provides an outlook on possible future research.

2   Swarm-Based Algorithm

Our swarm-based coverage and connectivity maintenance algorithm is based on the schooling behaviour of fish where fish usually of the same species, age and size stay together and swim in the same direction in a coordinated manner. The basic rules behind the schooling behaviour of fish are the same in all animal aggregation behaviours such as flocking of birds, the swarming of insects, and herding of land animals. Reynolds first modelled the flocking behaviour with his simulation program, Boids [8]. The basic model of flocking or schooling is controlled by the following three simple rules:

- **Separation:** A force which pushes neighbours away, increasing the size of the swarm.
- **Cohesion:** An attractive force to the average neighbour-position, maintaining the connectivity of the swarm.
- **Alignment:** Matching the average speed of the neighbourhood.

Formalizing the above three rules, the corresponding three forces $F_{separation}$, $F_{cohesion}$ and $F_{alignment}$ exerted on each member of the swarm is calculated as follows:

$$F_{separation} = \frac{1}{|neighbours|} \sum_{i \in neighbours} \frac{K}{(|\boldsymbol{p} - \boldsymbol{p_i}|)^2} \frac{\boldsymbol{p} - \boldsymbol{p_i}}{|\boldsymbol{p} - \boldsymbol{p_i}|} \tag{1}$$

$$F_{cohesion} = (\frac{1}{|neighbours|} \sum_{i \in neighbours} \boldsymbol{p_i}) - \boldsymbol{p} \tag{2}$$

$$F_{alignment} = \frac{1}{|neighbours|} \sum_{i \in neighbours} \boldsymbol{F_i} \tag{3}$$

where p is the position of the current member of the swarm, $\boldsymbol{p_i}$ is the position of its ith neighbour, $\boldsymbol{F_i}$ is the force of its ith neighbour and K is a *repulsion* parameter. The resulting force F_{result} on each robot is the weighted sum of the three forces $F_{separation}$, $F_{cohesion}$ and $F_{alignment}$,

$$F_{result} = w_1 \cdot F_{cohesion} + w_2 \cdot F_{separation} + w_3 \cdot F_{alignment} \tag{4}$$

where w_1, w_2 and w_3 denote the weights of the forces. The sum of the weights w_1, w_2 and w_3 is set to 1. Usually the weights w_1 is set equal to w_2 and w_3 to a smaller value compared to w_1 and w_2. Setting w_3 to a small value ensures that the alignment force gets lessened each time it spreads through the network and results in a more stable behaviour. It is also possible to set the alignment weight to zero.

A robot reaches a state of equilibrium when the total force exerted on it is zero. We design the repulsion parameter K in such as way that the state of equilibrium is achieved when the robot reaches an aimed distance D_{AIM} from all its neighbours. Let n be the number of neighbours of the current robot. At the aimed distance,

$$D_{AIM} : F_{cohesion} + F_{separation} = 0 \tag{5}$$

We calculate K as follows:

$$F_{cohesion} = (\frac{1}{n} \sum_{i \in n} \boldsymbol{p_i}) - \boldsymbol{p}$$

$$= \frac{1}{n}(\boldsymbol{p_1} + \boldsymbol{p_2} + \ldots + \boldsymbol{p_n}) - \boldsymbol{p}$$

$$= \frac{\boldsymbol{p_1} + \boldsymbol{p_2} + \ldots + \boldsymbol{p_n} - n\boldsymbol{p}}{n} \tag{6}$$

$$F_{separation} = \frac{1}{n} \sum_{i \in n} \frac{K}{(|\boldsymbol{p} - \boldsymbol{p_i}|)^2} \frac{\boldsymbol{p} - \boldsymbol{p_i}}{|\boldsymbol{p} - \boldsymbol{p_i}|}$$

$$= \frac{1}{n} \sum_{i \in n} \frac{K \cdot (\boldsymbol{p} - \boldsymbol{p_i})}{(|\boldsymbol{p} - \boldsymbol{p_i}|)^3}$$

$$= \frac{1}{n} \frac{K((\boldsymbol{p} - \boldsymbol{p_1}) + (\boldsymbol{p} - \boldsymbol{p_2}) + \ldots + (\boldsymbol{p} - \boldsymbol{p_n}))}{(|D_{AIM}|)^3}$$

$$= \frac{-1}{n} \frac{K(\boldsymbol{p_1} + \boldsymbol{p_2} + \ldots + \boldsymbol{p_n} - n\boldsymbol{p})}{(|D_{AIM}|)^3} \tag{7}$$

From equation 5, 6 and 7, we get   (8)

$$\frac{p_1 + p_2 + \ldots + p_n - np}{n} = \frac{1}{n} \frac{K(p_1 + p_2 + \ldots + p_n - np)}{(|D_{AIM}|)^3}$$

$$K = (|D_{AIM}|)^3 \qquad (9)$$

3 Experimental Analysis

We now compare the performance of our algorithm, referred to as *Swarm*, with the state of the art algorithms using the open source Player-Stage [1] robotic platform. The most prominent state-of-the art algorithms use force-based [4, 7, 10] and Voronoi-based [9] algorithms for coverage maximization. Force-based approaches consider robots as virtual particles driven by virtual forces. The most popular force based algorithms use artificial potential field-based forces [4, 5]. In [7], an extension of this approach to assure K-connectivity has been presented. Inspired by the equilibrium of molecules, a distributed self spreading algorithm *DSSA* has been presented in [2]. In [9], three Voronoi based algorithms VEC, VOR, Minimax has been proposed, which use the structure of Voronoi diagrams for finding coverage holes and minimizes them by relocating the robots.

We consider the force-based approach presented in [7] referred to as *Force*, *DSSA*, *VEC*, *VOR* and *Minimax* for comparing the performance of *Swarm*. For the *Swarm* algorithm, we find the net force on each robot according to equation 4. The implementations of *Force* and *DSSA* algorithms are straight forward. For *Force* implementation, we find the constants K_{cover} and K_{degree} exactly as mentioned in [7] and assign the same values, 0.25 and 0.8, used in the paper for the damping and safety factors. For *DSSA*, the expected density μ and the local density D are calculated as mentioned in [2].

During the implementation of Voronoi based algorithms, the following issue arises: If robots are cluttered together in a small area, the robots at the border of the cluttered area might construct open Voronoi polygons. Checking the existing Voronoi points cannot detect coverage holes in such cases and the coverage is not improved. To solve this issue, we assume that the area where the robots are deployed has known boundaries. When the Voronoi polygons are open or the existing Voronoi points are located outside the boundary, new Voronoi points from the boundary are added. This makes the polygons closed within the defined area and moves the robots from cluttered regions towards sparser regions. We use this extended approach in our experiments.

We assume isotropic radial sensors of range r_s and isotropic radio communication of range r_c for our robots, with $r_c \geq 2 \cdot r_s$. We set r_c to 1.0, r_s to 0.5 and D_{AIM} to $\sqrt{3} \cdot r_s$. The experiment is conducted on a square field of size 6×6 meters in an obstacle free area. 55 robots are deployed randomly in the field. We now evaluate the performance of the algorithms based on the rate of coverage, which indicates the increase/decrease in coverage over time and the total final coverage achieved.

Figure 1 shows the performance of these six algorithms averaged over 50 independent runs with different random initial configurations. It shows that the

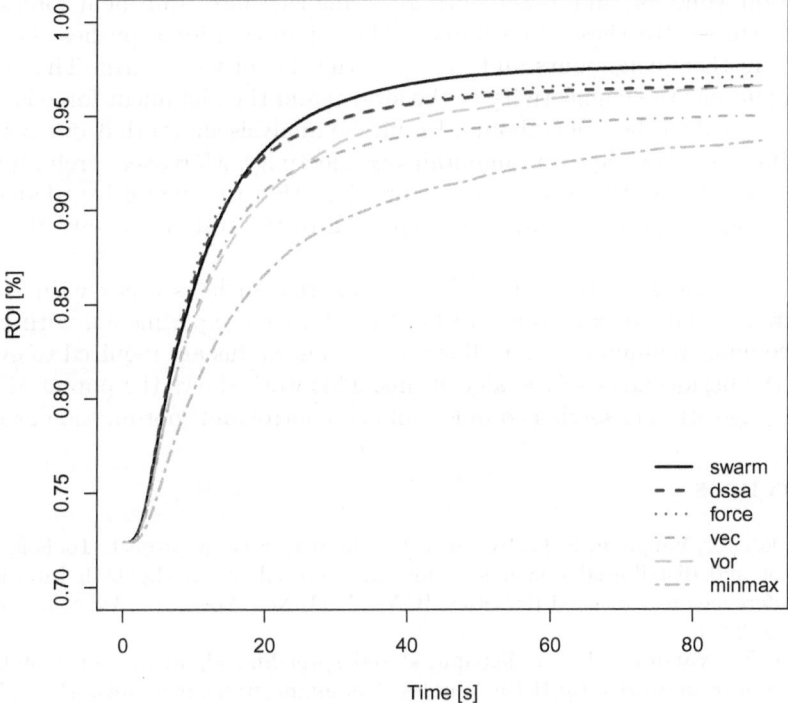

Fig. 1. Comparison of *Swarm* algorithm's performance with that of *Force*, *DSSA*, *VEC*, *VOR* and *Minimax* algorithms

Swarm algorithm performs better than state of the art algorithms by covering more area in lesser time. This is achieved with the three forces used in *Swarm* algorithm, where separation and cohesion forces help the robots to quickly attain D_{AIM} distance from their neighbours and alignment force helps them to relocate faster while overcoming coverage holes.

Minimax and VOR algorithms are expected to perform better than VEC algorithm. However, in our experiments VOR achieves better coverage than VEC algorithm only with relatively longer time. The coverage of Minimax is much lesser than VOR and VEC. One reason for this variation could be the difference in the simulator, where the robots do not move to the Minimax point in each iteration of the algorithm due to the randomness and unsynchronized execution of the algorithms in the Stage simulator.

4   Conclusion and Future Work

In this paper, we focussed on the design of a new bio-inspired algorithm for solving the coverage and connectivity maintenance problem. Inspired from nature, we designed our swarm-based algorithm based on the schooling behaviour of fish. In this algorithm, the three basic rules behind the schooling behaviour, separation, cohesion and alignment, are formalized into equivalent forces. Each robot is subject to these three forces. The separation force pushes the robots away from their neighbours and increases the size of the swarm. The cohesion force maintains the connectivity of the swarm and the alignment force keeps the robots aligned to their neighbours. Empirical analysis shows that our new algorithm improves coverage and maintains connectivity. Moreover, preliminary results obtained from the basic experiments show that the swarm-based algorithm achieves better and faster coverage compared to the state-of-the-art algorithms tested.

More empirical analyses in different scenarios such as areas with complex contours and with obstacles are to be done. Further experiments, with varying robot counts, communication radius and sensing radius are required to evaluate better the performance of the algorithms. This work shows the potential of bio-inspired algorithm in solving complex robotic control and coordination problems.

References

1. Gerkey, B., Vaughan, R.T., Howard, A.: The player/stage project: Tools for multi-robot and distributed sensor systems. In: Proceedings of the 11th International Conference on Advanced Robotics, ICAR 2003, New York, NY, USA, pp. 317–323 (June 2003)
2. Heo, N., Varshney, P.: A distributed self spreading algorithm for mobile wireless sensor networks. In: IEEE Wireless Communications and Networking, WCNC 2003, vol. 3, pp. 1597–1602 (March 2003)
3. Howard, A., Matarić, M.J., Sukhatme, G.S.: An incremental self-deployment algorithm for mobile sensor networks. Autonomous Robots Special Issue on Intelligent Embedded Systems 13(2), 113–126 (2002)
4. Howard, A., Mataric, M.J., Sukhatme, G.S.: Mobile sensor network deployment using potential fields: A distributed, scalable solution to the area coverage problem. In: DARS: 6th International Symposium on Distributed Autonomous Robotics Systems, pp. 299–308 (June 2002)
5. Khatib, O.: Real-time obstacle avoidance for manipulators and mobile robots. In: 1985 Proceedings of the IEEE International Conference on Robotics and Automation, vol. 2, pp. 500–505 (March 1985)
6. McLurkin, J., Smith, J.: Distributed algorithms for dispersion in indoor environments using a swarm of autonomous mobile robots. In: 7th International Symposium on Distributed Autonomous Robotic Systems, DARS (2004)
7. Poduri, S., Sukhatme, G.S.: Constrained coverage for mobile sensor networks. In: IEEE International Conference on Robotics and Automation, New Orleans, LA, pp. 165–172 (May 2004)
8. Reynolds, C.W.: Flocks, herds and schools: A distributed behavioral model. SIGGRAPH Computer Graphics 21, 25–34 (1987)
9. Wang, G., Cao, G., La Porta, T.F.: Movement-assisted sensor deployment. IEEE Transactions on Mobile Computing 5, 640–652 (2006)
10. Zou, Y., Chakrabarty, K.: Sensor deployment and target localization based on virtual forces. In: Twenty-Second Annual Joint Conference of the IEEE Computer and Communications, INFOCOM 2003, March 30-April 3, vol. 2, pp. 1293–1303. IEEE Societies (2003)

Monitoring of a Large Wi-Fi Hotspots Network: Performance Investigation of Soft Computing Techniques[*]

Pheeha Machaka, Takudzwa Mabande, and Antoine Bagula

Intelligent Systems and Advanced Telecommunication Laboratory (ISAT)
Department of Computer Science, Room 317 Computer Science Building, 18 University
Avenue, University of Cape Town, Rondebosch, Cape Town, South Africa, 7701
{pheeha.machaka,takudzwa.mabande,antoine.bagula}@uct.ac.za

Abstract. This paper addresses the problem of network monitoring by investigating the performance of three soft computing techniques, the Artificial Neural Network, Bayesian Network and the Artificial Immune System. The techniques were used for achieving situation recognition and monitoring in a large network of Wi-Fi hotspots as part of a highly scalable preemptive monitoring tool for wireless networks. Using a set of data extracted from a live network of Wi-Fi hotspots managed by an ISP, we integrated algorithms into a data collection system to detect anomalous performance and aberrant behavior in the ISP's network. The results are therefore revealed and discussed in terms of both anomaly performance and aberrant behavior on several test case scenarios.

Keywords: Performance Monitoring, Neural Networks, Artificial Immune Systems, Bayesian Networks, Anomaly Performance Detection, Aberrant Performance Detection.

1 Introduction

The popularity of Wireless Fidelity (Wi-Fi) technology has led to a large scale deployment of thousands of hotspots networks. Wi-Fi is based on the family of IEEE 802.11 standards and builds upon a fast, easy and inexpensive approach to networking [1] where Access Points (APs) are used to broadcast signals to Wi-Fi-capable client devices (laptops and Smartphone devices) within their range, and connect to the Internet. Performance monitoring is an important task upon which large Wi-Fi network deployment depends. Wi-Fi hotspots generate massive monitoring datasets which require efficient performance monitoring methods to analyze data, recognize anomalous hidden patterns and implement fault tolerance. As traditionally implemented, performance monitoring in Wi-Fi networks is based on a reactive network approach where the operating system software only warns the network

---

[*] The financial assistance of the National Research Foundation (NRF) and Telkom SA Center of Excellence (CoE) is hereby acknowledged. Opinions expressed and conclusions arrived at, are those of the author and are not necessarily to be attributed to the NRF and CoE.

E. Hart et al. (Eds.): BIONETICS 2011, LNICST 103, pp. 155–162, 2012.
© Institute for Computer Sciences, Social Informatics and Telecommunications Engineering 2012

administrators when a problem occurs. This approach leads to both the halting of important network processes and the hampering of critical business processes of the organization. Pre-emptive network monitoring provides the potential to prevent the occurrence of faults by analyzing the status of the network components, to create a fail-safe network status or allow a smooth migration from a faulty to fail-safe network status. While statistical analysis has been deployed in many cases to address this issue, soft computing methods are emerging as powerful tools used in anomaly detection and security monitoring systems.

1.1   Related Work

There has been large amounts of work done in the field of anomaly detection, and in this paper, three soft computing methods were identified, viz. Artificial Neural Networks, Artificial Immune Systems and Bayesian Networks. With Artificial Neural Network (ANN), the work has focused on employing ANN for anomaly detection on network traffic data [2-4]. Artificial Immune Systems (AIS) was used for intrusion detection, and detection of computer viruses[5-7].Bayesian Networks were also used for anomaly detection for disease outbreak [4] and also in detecting and analyzing anomaly behavior in network-based FTP services [8].The three methods have gained success in anomaly detection and in this paper, we would like to employ them on a large network of Wi-Fi hotspots and answer the following questions:

1. Which method performs better?
2. How do they perform under different test cases and network thresholds?
3. How do the methods perform for anomaly detection, and aberrant behavior detection?

The remainder of this paper is organized as follows. Section 2 describes the algorithms used for the experiments in this paper. Section 3 describes experiment's model, performance metrics and test cases for the experiment. In section 4 and 5, the results of the experiment are revealed and discussed.

2   Algorithms

2.1   Artificial Neural Networks

ANN's are mathematical or computational models that get their inspiration from biological neural systems. In this paper the neural network model, Multilayer Perceptron (MLP) was used to conduct experiments. The MLP is a feed forward neural network model in which vertices are arranged in layers. MLP have one or more layer(s) of hidden nodes, which are not directly connected to the input and output nodes [9]. For the purpose of this experiment we employed Weka's Multilayer Perceptron implementation.

2.2   Bayesian Networks

Bayesian Networks can be described briefly as acyclic directed graph (DAG) which defines a factorisation of a joint probability distribution over the variables that are represented by the nodes of the DAG, where the factorisation is given by the direct links of the DAG [10]. The NaiveBayes algorithm was used for the experiments. It makes a strong assumption that all attributes of the examples are independent of each other given the context of the class. The Weka's NaiveBayes implements this probabilistic Naïve Bayes classifier [11].

2.3   Artificial Immune System

The AIS takes inspiration from the robust and powerful capabilities of the Human Immune System's (HIS) capabilities to distinguish between self and non-self [7].The Algorithm employed in this paper's experiments is the Weka's Artificial Immune Recognition System (AIRS) learning algorithm [12].The AIRS is a supervised AIS learning algorithm that has shown significant success on a broad range of classification problems [5-7].

3   The Experimental Model

3.1   The Wi-Fi Network

The experiment network was made of more than 400 hotspots around the Cape Town area, with more than 615 Cisco WRT54GL gateway devices connected to the network. A Syslog daemon program installed on each gateway device, and was left to run from 2009-07-03 12h00 to 2009-09-02 04h00 collecting monitoring data at every hour's interval, thus leading to up to 356537 items in the experiment dataset.

3.2   Performance Metrics, Encoding and Selection

Monitoring. The following three performance metrics were monitored in the network:

- Uptime and Downtime (%) - This metric measures the availability, stability and reliability of the communication device when used in the network.
- Load Average (%) - Measures the "congestion rate" for the device based on the number of users connected to the device.
- Radio Noise (in dB) - Wi-Fi uses the shared 2.4 GHz spectrum band and the proliferation of devices using the spectrum leads to congestion and noisy Wi-Fi devices.
- Standard deviation - To detect aberrant behavior in performance, statistical confidence bands were used to measure deviations in a time series. A deviation depended on the Delta (δ) parameter whose sensible values were taken between 2 and 3 [13].

Encoding and Selection. Three levels of performance were used to describe performance. A 3-bit encoded nominal value was used to describe performance using the test case shown in table 1.

1 $\rightarrow$ LOW LEVELS, 2$\rightarrow$MODERATE LEVELS, 3$\rightarrow$ HIGH LEVELS

3.3   Performance Evaluation of the Methods

The effectiveness of the methods is evaluated based on their ability to make correct predictions. The following measures were used to quantify the performance of the algorithms:

- True Positive (TP) rate, also known as detection rate.
- False Positive (FP) rate, also known as false alarm rate.
- F-measure, it is a harmonic mean for precision and recall.
- Kappa Statistic - used to measure the agreement between predicted and observed categorization of a dataset, while correcting for agreement that occurs by chance.

3.4   Test Cases

We conducted experiments using four test case scenarios revealing Wi-Fi operating constraints from loose (e.g. rural setting where QoS is an issue) to the most stringent(e.g. suburban setting where modern applications demand QoS). In the experiments conducted, four sets of test cases were devised. These test cases are defined by Table 1 in terms of Noise, Load, Downtime rate (overall time the device was down) and the confidence band δ.

Using the dataset from the data collection component, a stratified 10-fold cross-validation technique was used for training and testing the algorithms.

4   Experimental Results

Using the parameters described in the test cases of Table 1, we conducted a set of experiments to detect anomalous and aberrant network performance.

4.1   Anomaly Performance Evaluation

This section will report on the anomaly performance results for the experiment per performance evaluation method. Anomaly performance refers to those measurements that perform above a given threshold level.

Anomaly True Positive Rate. Figure 1 indicates that the MLP had an average TP rate of 99.45%, while NaiveBayes had an average TP rate of 95.62% across all test cases. NaiveBayes and MLP were able to correctly identify the most faults correctly in all test cases. The AIRS1was the poorest technique in recognising classes correctly with average TP rate of 47.65%.

Anomaly False Positive Rate. The MLP and NaiveBayes had very low average FP rate, 0.77% and 6.45% respectively. This is indicated in figure 1. A poor performance was seen with AIRS1 technique, it had an average FP rate of 23.65%, and had a high FP rate of 38.2% in test case 4.

Anomaly F-Measure. The MLP is shown to be on average, the most accurate of the techniques with an average F-measure of 99.45%. The NaiveBayes had an average

Table 1. Parameter settings for each experiment test case

Test Cases	Noise(x100 dB)		Load (%)	Downtime (%)	Confidence Band
1	high	-7000	100		1.00
	moderate	-8000	70	4%	2.00
	low	-9000	0		3.00
2	high	-7250	100		0.75
	moderate	-8250	75	3%	1.75
	low	-9250	0		2.75
3	high	-7500	100		0.50
	moderate	-8500	80	2%	1.50
	low	-9500	0		2.50
4	high	-7750	100		0.25
	moderate	-8750	85	1%	1.25
	low	-9750	0		2.25

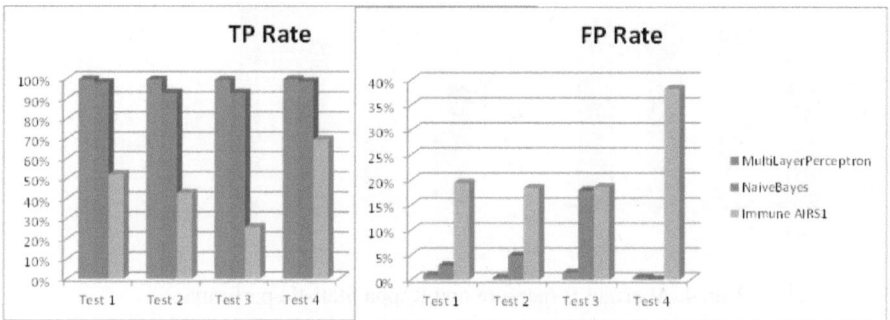

Fig. 1. Anomaly True positive and False Positive rate performance

F-measure of 95.25% across all test cases. This is indicated by figure 2. AIRS1 revealed poor results with an average F-measure of 53.88%.

Anomaly Kappa Statistic. Shown in figure 2, the MLP and NaiveBayes had an average Kappa statistic of 98.59% and 89.05%, respectively. AIRS1 technique had an average Kappa statistic of 15.22%, revealing poor accuracy and precision.

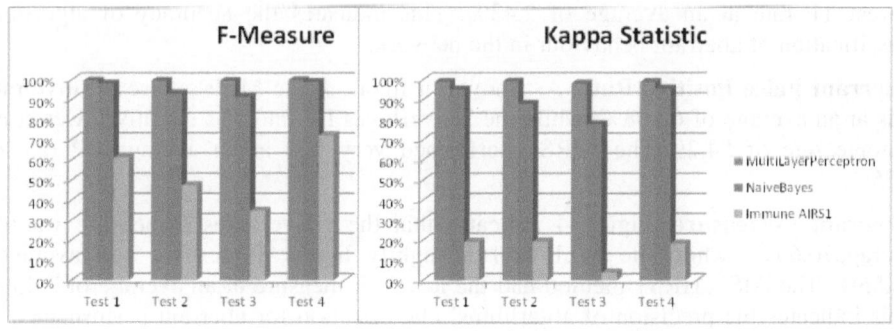

Fig. 2. Anomaly F-Measure performance and Kappa Statistic

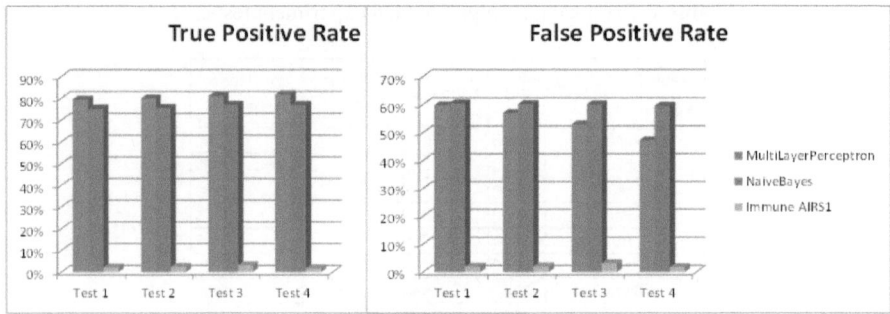

Fig. 3. Aberrant True Positive and False Positive rate performance

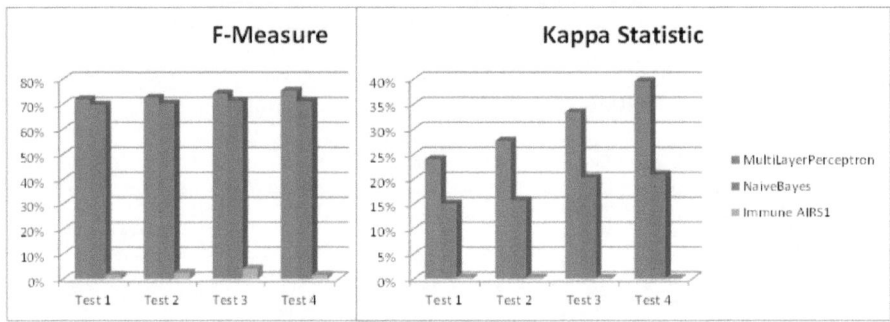

Fig. 4. Aberrant F-measure and Kappa Statistic performance

4.2   Aberrant Performance Evaluation

This section of the paper reports on detection of aberrant behavior. An observation will fall into the aberrant behavior category when it's observed values fall outside the statistical confidence band, an outlier. A total deviation of the observation will depend on a defined Delta (δ) parameter.

Aberrant True Positive Rate. Shown in figure 3, MLP's average TP rate was at 76.2%. The NaiveBayes had a slightly higher TP rate of 80.6%. The AIRS1 had the lowest TP rate at an average of 2.43%. This indicates the accuracy of algorithm classification of aberrant behaviour in the network.

Aberrant False Positive Rate. As Shown in figure 3, the MLP's false positive rate was at an average of 60.08%, while the NaiveBayes FP rate was slightly lower at an average rate of 54.2%.The AIRS1 performed very low at an average FP rate of 2.25%.

Aberrant F-Measure. Figure 4 indicates that the NaiveBayes F-measure was on average70.65%, while the MLP had a slightly higher F-measure and averaged 73.55%. The AIS' AIRS1 method had the lowest F-measure at an average of 2.45%. This indicates the precision of algorithms' classification for aberrant performance in the network.

Aberrant Kappa Statistic. The AIRS1 had a poor Kappa statistic with an average of 0.23% across all test cases. The NaiveBayes method had a slightly higher Kappa Statistic which averaged at 18.00% across all test cases. The MLP had a higher and fair Kappa statistic which averaged 31.16% across all test cases. The MLP had a higher Kappa Statistic in comparison with the NaiveBayes and AIRS1 methods. Although its kappa statistic was higher, it was not anywhere close to perfect, but fair.

5   Conclusion

Aiming to achieve pre-emptive monitoring of wireless networks, this paper proposes computational algorithms which use anomalous and aberrant behaviors to measure the performance of a large Wi-Fi network. Anomalous performance refers to network devices whose performance is outside or above a desired performance threshold. The artificial immune neural network's multilayer perceptron (MLP) and the NaiveBayes algorithms managed to classify the network's anomalous performance with higher accuracy and precision as shown by their true positive rate; false positive rate; F-measure and Kappa statistic performance. On the other hand, the artificial immune system's AIRS1 displayed poor classification of anomalous network performance.

Aberrant performance refers to network devices that experienced highly variable and unstable changes in performance. When looking at aberrant behavior, we found that the AIRS1's performance was neither satisfactory nor convincing. On the other hand, the performance revealed by both MLP and NaiveBayes in terms of true positive rate, F-measure and Kappa statistic were good and fair, but they had an alarming high false positive rate performance.

The results obtained by the experiments conducted in this research reveal that both MLP and NaiveBayes are better suited for anomaly performance detection, and MLP showed a slightly higher accuracy and precision. One would need to be careful when implementing both algorithms for aberrant performance detection. More research and experiments may be required to improve accuracy and precision for aberrant performance detection.

References

1. Vaughan-Nichols, S.J.: The challenge of wi-fi roaming. Computer 36, 17–19 (2003)
2. Cannady, J.: Artificial neural networks for misuse detection. In: Proceedings of the 21st National Information Systems Security Conference, Arlington, VA, USA (1998)
3. Cheng, E., Jin, H., Han, Z., Sun, J.: Network-Based Anomaly Detection Using an Elman Network. In: Lu, X., Zhao, W. (eds.) ICCNMC 2005. LNCS, vol. 3619, pp. 471–480. Springer, Heidelberg (2005)
4. Zhang, J., Zulkernine, M.: Anomaly based network intrusion detection with unsupervised outlier detection. In: ICC 2006, Istanbul, vol. 9 (2006)
5. Forrest, S., Hofmeyr, S.A., Somayaji, A., Longstaff, T.A.: A sense of self for unix processes. In: 1996 Proceedings of IEEE Symposium on Security and Privacy (1996)
6. Dasgupta, D., Gonzalez, F.: An immunity-based technique to characterize intrusions in computer networks. IEEE Transactions on Evolutionary Computation 6, 281–291 (2002)

7. Luther, K., Bye, R., Alpcan, T., Muller, A., Albayrak, S.: A cooperative AIS framework for intrusion detection. In: 2007 IEEE International Conference on Communications (2007)

8. Cha, B., Lee, D.: Network-Based Anomaly Intrusion Detection Improvement by Bayesian Network and Indirect Relation. In: Apolloni, B., Howlett, R.J., Jain, L. (eds.) KES 2007, Part II. LNCS (LNAI), vol. 4693, pp. 141–148. Springer, Heidelberg (2007)

9. Dunne, R.A.: A statistical Approach to Neural Network for Pattern Recognition, p. 288. Wiley-Interscience (2007)

10. Kjaerulff, U.B., Madsen, A.L.: Bayesian Networks and Influence Diagrams: Guide To Construction and Analysis, 1st edn., p. 336. Springer (2007)

11. Witten, I.H., Frank, E.: Data Mining: Practical Machine Learning Tools and Techniques, 2nd edn. Morgan Kaufmann Series in Data Management Systems, p. 560. Morgan Kaufmann (2005)

12. Watkins, A., Timmis, J., Boggess, L.: Artificial immune recognition system (AIRS): An immune-inspired supervised learning algorithm. Genetic Programming and Evolvable Machines 5, 291–317 (2004)

13. Robinson, S.: Simulation: The Practice of Model Development and Use, 1st edn., p. 339. John Wiley & Sons Ltd., The Atrium (2004)

Gossip Inspired Sensor Activation Protocol for a Correlated Chemical Environment

Champake Mendis

Computer Science and Software Engineering,
The University of Melbourne, Carlton, VIC, 3053 Australia
mendisc@csse.unimelb.edu.au

Abstract. The energy conservation in chemical sensor networks is crucial as chemical sensors with air sampling to consume significant energy for sensing activity compared to that used for communication unlike other types of sensors, such as optical or acoustic. When considering the threat environment, the chemical tracers dispersed by turbulent motion in the environment display rather complex and even chaotic properties. Hazardous chemical releases are rare events. If all sensors in a wireless chemical sensor network (WCSN) are left in the active state continuously, it will result in significant power consumption. Therefore, dynamic sensor activation is essential for the durability of WCSNs. Dynamic sensor activation for chemical sensor networks using an epidemiology-based sensor activation protocol has been proposed in the literature. In this paper, we investigate the performance of a variant of epidemiology, gossip inspired sensor activation protocol of a WCSN in a chemical tracer field. The simulation framework with gossip protocol is validated against an analytical model. We then perform simulation experiments to evaluate the performance of gossip- based sensor activation protocol on selected performance metrics: number of active sensors and reliability of detection. We show by simulations that by varying the communication radii of sensors, we can achieve better energy conservation while maintaining better performance of a WCSN with a gossip-based activation protocol.

Keywords: Wireless chemical sensor network, bio-inspired algorithms, gossip algorithm, epidemiological model, chemical tracers.

1 Introduction

A chemical detection system is comprised of individual sensors or a network of sensors and an environment prone to a possible threat from chemicals. The chemical sensors have unique characteristics that distinguish them from other types of sensors that are typically used in wireless sensor networks. Particularly, unlike other types of sensors, such as optical or acoustic sensors, typical chemical sensors with air sampling to consume significant energy for sensing activity compared to what is consumed for communication. When considering the threat environment, the chemical tracers dispersed by turbulent motion in the environment display rather complex and even chaotic properties (so-called phenomenon of scalar turbulence [1]). Because hazardous chemical releases are rare events, if all sensors in a WCSN are left in the active state continuously, it would result in significant power consumption. Therefore, dynamic sensor

E. Hart et al. (Eds.): BIONETICS 2011, LNICST 103, pp. 163–170, 2012.

activation is crucial for the longevity of WCSNs. Dynamic sensor activation for chemical sensor networks using an 'epidemiological' sensor activation protocol has been proposed in [2], [3] and [4]. In [4], the authors found that random network topology and spatial correlation of tracer distribution (which is typical for most operational scenarios of chemical environments) to have a negative effect on the performance of a WCSN with the epidemiology based sensor activation protocol and they have studied scalability factors as well [4]. Here, we introduce a randomised 'gossip' factor to the existing epidemiological protocol to retain the simplicity of expressing its behaviour by an analytical expression. When the gossip factor reaches unity, we show that our protocol behaves similar to epidemiology-based protocol in [4]. When gossip factor is zero, network becomes dysfunctional. In compliance with our intuitive assumption that gossip-based protocol will solve the sensor activation anomaly in a correlated chemical environment by extending the communication radius while reducing the number of active sensors, we perform simulation experiments to study the behaviour of gossip-based protocol on a chemical environment using selected performance metrics: number of active sensors and reliability of detection which are well-established metrics for other types of sensor networks (see [5]).

The paper is organised as follows. Section 2 discusses related work. Section 3 describes system modelling and simulation. In Section 4, we describe the analytical model, and in Section 5, we validate the simulation model. Evaluation of WCSN is presented in Section 6. Finally, in Section 7, we draw conclusions and suggest future work.

2   Related Work

The gossip algorithm was initially introduced by Demers *et al.* in [6]. Eugster *et al.* [7] have studied the factors such as membership management in message transmission associated with gossip algorithm application. Rao *et al.* [8] have applied the gossip algorithm to peer-to-peer networking to extract aggregated information from the network. The studies of Haas *et al.* [9] and Boyd *et al.* [9] are instances for application of gossip protocol to wireless sensor networks. In the literature, the use of gossip algorithm is grouped into three types. (i.e., information spread, aggregate computation, overlay management), and we found that the gossip protocol is expressed in complex mathematical forms and interpretation is difficult. In physical implementations, Gamm *et al.* [10] present a sensor with two power modes (i.e., sleeping and receiving, transmission). This substantiates the feasibility of using a gossip-based protocol in sensor activation with existing sensor technology. In [4], substantial studies have been done on correlated and non-correlated chemical tracer fields, random and regular grid topologies, and effect of WCSN design factors. In this paper, we present the findings of a simulation study performed to evaluate the performance of a WCSN with a gossip-based sensor activation protocol in the presence of chemical tracer fields. Specifically, we make the following contributions:

- We propose a simple analytical model for gossiping algorithm by introducing a gossiping factor to an epidemiological model.
- We evaluate the behaviour of a WCSN in complex chemical tracer fields on some performance metrics of the WCSN (namely, number of active sensors and reliability of detection) to obtain optimal network sensor configurations.

We believe that these contributions sufficiently illustrate the novelty of the present study.

3 System Modelling and Simulation

For the purpose of simulating WCSN with a gossip-based sensor activation protocol, we modified the simulation framework used in [4]. It comprises three independent models representing the environment (the spatiotemporal realisation of the concentration field of the chemical tracers caused by turbulent mixing), an individual chemical sensor node, and the whole WCSN.

We consider a chemical tracer field covering a square region of side length L and area S ($S = L^2$). We assume that all sensor nodes of our system are deployed at the same height; so, we are interested in 2D tracer concentrations. The chemical tracer field is modelled as a sequence of time-varying random data slices, which mimics the turbulent fluctuations of concentration at each point in the 2D xy-plane of interest where the sensor nodes are deployed [11]. The mean concentration of the tracer field is C_0.

We consider each sensor node to have a chemical tracer detection threshold C_* and a detection period τ_*. The aggregated value of concentration detected by active sensors is C. The sensors are capable of communicating in an omni-directional fashion with its neighbouring nodes that are within a communication range r_*. By normalising the communication radius r_* by tracer field size L, it is expressed as a non-dimensional parameter δ as follows [4]:

$$\delta = \frac{r_*}{L}. \tag{1}$$

We consider a WCSN consisting of N identical chemical sensor nodes and covering the full area of the chemical tracer field, and WCSN will have a response time τ. We accept that energy consumed when sensor is activated is greater than extending the communication radius of a sensor based on the energy measurements specified in [10].

4 The Analytical Model

The overall behaviour of the WCSN is similar to human gossiping and analogous to an epidemic SIS (susceptible-infected-susceptible) model in [12]. Based on this intimate analogy with the process of an epidemic, an analytical model in the form of a closed system of ODE can be derived to describe the dynamics of the WCSN [2]:

$$\frac{dN_+}{dt} = \alpha N_+ N_- - \frac{N_+}{\tau_*}, \tag{2}$$

$$\frac{dN_-}{dt} = -\alpha N_+ N_- + \frac{N_+}{\tau_*}, \tag{3}$$

where N_+ and N_- denote the number of active and passive sensors, respectively. The population size (i.e., the total number of sensors) is conserved, that is $N_+ + N_- = N = const$. The non-linear terms on the RHS of Eq. (2) and Eq. (3) are responsible for the interaction between individuals (i.e., sensors), with the parameter α being a measure of this interaction. Based on the epidemiological analogy, the following analytical expression for the interaction parameter α is proposed by Skvortsov *et al.* [2]:

$$\alpha = G\frac{\pi r_*^2}{\tau_* S}p, \tag{4}$$

where S is the square area covered by a WCSN, and G is the calibration factor, being of order unity, and parameter p, as above, is expressed in terms of the cumulative distribution function: $p = 1 - F(C_*)$; parameter p is a function of the sensor detection threshold C_* and correlation parameter ω. In the synthetic chemical tracer model we use, $C_* < C_0$, p increases with ω, and when $C_* > C_0$, p decreases with ω; when $C_* = C_0$, p is independent of ω [4].

We can express the interaction rate of gossip-based WCSN α_G as a function of α given in Eq. (4):

$$\alpha_G = \alpha f, \tag{5}$$

where f is the *fanout* ratio [7] or gossiping probability [9] of gossip-based sensor activation protocol. The epidemiological protocol, represented by Eq. (2) and Eq. (3), can be easily treated analytically as the gossip protocol with Eq. (5) and allows us to model a variety of responses and perform various optimisation studies. Skvortsov *et al.* [2] found the analytical solution of the system to be given by:

$$z(t) = \frac{z_0}{(1 - z_0)\exp(-bt) + z_0}, \tag{6}$$

where $z(t) = N_+/N$, $z_0 = z(0)$, and $b = \alpha N - \frac{1}{\tau_*}$.

This analytical solution provides us with a valuable performance check for our simulations. When $f = 1$, WCSN overall behaviour becomes epidemiological-based, and when $f = 0$, WCSN becomes dysfunctional. In our simulations, we predefine the activation pattern of all sensors-based on f, where we deviate from the decentralized sensor activation approach proposed in [4]. For implementation, we use a vectorised form to select sensors synchronously for activation to improve the simulation speed. In real physical implementation, this can be accepted as a 'black list' embedded in a message sent by a sensor to activate its neighbour after detecting a chemical tracer concentration.

5   Validation of WCSN Model

To validate our WCSN simulation model against the analytical model of Skvortsov *et al.* [2] described in Section 4, the simulation experiments were performed using WCSNs of regular grid topology, with a chemical tracer field [4]. The following simulation parameter values were used for the WCSN in these validation experiments: $N = 400$ sensors, $L = 250$ units, and $r_* = 20$ units (equivalently, $\delta = 0.08$ in the non-dimensional form (1)) and $f = 0$, $f = 0.6$ and $f = 1$. All results in the paper will be presented in terms of non-dimensional parameters to make it possible to scale the results without being tied to particular measurement units.

Figure 1 shows the results of validation experiments for the WCSN model; in this figure, elapsed time t is shown as a non-dimensional parameter $\kappa = t/\tau_*$. The time evolution of the fraction of active sensors (N_+/N) obtained from simulations show sigmoidal patterns similar to that obtained by Eq. (6). We empirically estimated corresponding calibration constant G by fitting simulated data to the analytical equation. Here we measure the onset of *information epidemic* (IE) with fraction of active sensors N_+/N.

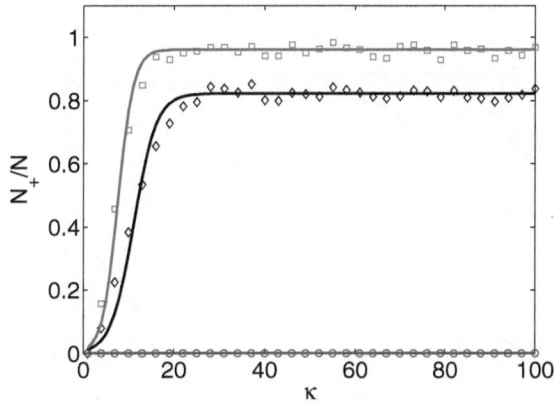

Fig. 1. The effect of fanout ratio (f) on the onset of IEs (N_+/N) in a WCSN with a regular grid topology in the presence of a chemical tracer field. The range of values used for the f are: $f = 0.0$ ($\circ$), $f = 0.6$ ($\diamond$), $f = 1.0$ ($\square$). The analytical fitted curves are shown in solid lines. $\kappa = t/\tau_*$ is the normalised time where τ_* is the detection period of a sensor.

As shown in Fig. 1, the good agreement between analytical and simulated results validates our simulation model. We see a rise in the saturation level of the IE and a decrease of response time τ when fanout ratio f is increased.

6   Evaluation

Describing the WCSN behaviour with analytical expressions provides a valuable insight into maximising the networking information gain and network performance optimisation. The simulation allows us to mimic a behaviour of a real system designed based on analytical derivations. In this section, we describe and present the results of two exercises performed to optimise the performance of WCSNs. In these, simulations were carried out using WCSN models in the presence of simulated chemical tracer field. Here, we use a graphical method to deduce optimal sensor and network configuration parameters.

6.1   Number of Active Sensors

The effect of fanout ratio on the sensor network's performance with different sensor communication ranges r_* was examined in this experiment. WCSNs of regular grid topology were simulated in the presence of a simulated chemical tracer field. The number of sensors was set at $N = 400$, and the sensor detection threshold was held at $C_* = C_0$. Simulations were performed for several values of communication radius r_* (or equivalently, the non-dimensional parameter δ); specifically, the following δ values were used: $\delta = 0.052, \delta = 0.056, \delta = 0.080, \delta = 0.160$, and $\delta = 0.480$ with 26 values of f ranging from 0 to 1.0. Each scenario was simulated 100 times and the ensemble-averaged results for sensor activation are shown in Figure 2.

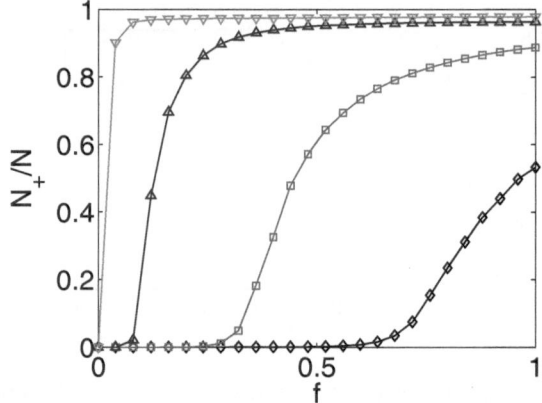

Fig. 2. The effect of fanout ratio (f) on the saturation value of (N_+/N) with different communication ranges (r_*) in a WCSN with a regular grid topology in the presence of a chemical tracer field. The range of values used for the non-dimensional communication radius parameter δ, defined in Eq. (1), are: $\delta = 0.056$ ($\diamond$), $\delta = 0.080$ ($\square$), $\delta = 0.160$ ($\triangle$), $\delta = 0.480$ ($\triangledown$).

We see a rise in the saturation level of the IE (N_+/N) when the communication radius r_* (or δ) is increased or when the fanout ratio (f) is increased. This can be explained with the analytical expression obtained in [2] modified and described by Eq. (6). We also observe that some low values of δ and f do not support an IE; i.e., the IE 'dies out', as the condition for an IE is not satisfied in these cases.

6.2   The Reliability of Detection

Here, we evaluate the reliability of detection; the results are shown in Figure 3. If we characterise the reliability of detection as the coefficient of variation of concentration detection C by active sensors $\Psi = \sigma_C/\mu_C$, where μ_C is the time-averaged mean of C (i.e., around the saturation level), and σ_C is the time-averaged fluctuations of C. Plots of Ψ versus f corresponding to several values of communication ranges δ are shown in Figure 3 for the WCSN.

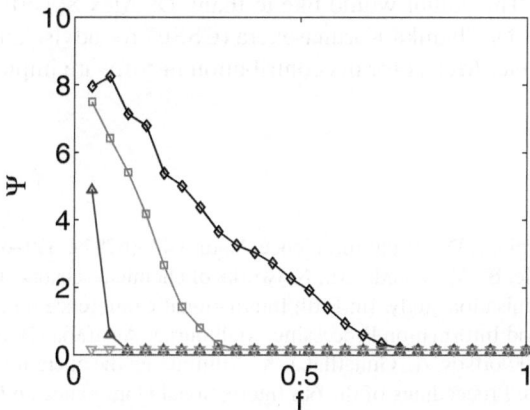

Fig. 3. The effect of fanout ratio (f) and communication radius (r_*) on the reliability of chemical tracer detection by a WCSN with a regular grid topology in the presence of a chemical tracer field. The values used for the non-dimensional communication radius parameter δ, defined in Eq. (1), are: $\delta = 0.056$ ($\diamond$), $\delta = 0.080$ ($\square$), $\delta = 0.160$ ($\triangle$), $\delta = 0.480$ ($\triangledown$).

From these plots, we can see that even when we decrease f by increasing r_*, we can obtain reliable readings from WCSN, and we can estimate the optimal points where Ψ reaches the zero or no fluctuations. Depending on a particular operational scenario, some other parameters (energy conservation, WCSN activation time, sensor interaction rate, budget, etc.) may become more important and would favour a particular technical solution.

7 Conclusions

In this paper, we investigated the behaviour of a WCSN with a bio-inspired gossip-based sensor activation protocol in a turbulent chemical tracer field, which is highly complex in nature. Using a bio-inspired analytical model for a WCSN, initially introduced by Skvortsov *et al.* [2], we verified the simulation framework for WCSNs (Fig. 1). Based on simulation results, we evaluated some important performance metrics: number of active sensors as shown in Fig. 2 and reliability of detection as shown in Fig. 3. Based on our results, we can consider extending the communication range of sensors than activating more neighbouring sensors, which is an energy conserving strategy [10]. The gossip protocol equips sensors with intelligence when selecting neighbour sensors for activation. This system can also be applied to a real system where sensor communications links get dysfunctional with time due to aging. In this, we used centralised aggregation of detected concentration values; we hope to incorporate gossip-based averaging [8] and bio-inspired optimisation technique to optimise the WCSN configuration parameters [3] in our future studies.

Acknowledgement. The author would like to thank Dr Alex Skvortsov, Dr Ajith Gunatilaka (DSTO), and Dr Shanika Karunasekera (CSSE) for advise and helpful technical discussions and Peter Morris for his contribution in software implementation of the initial model.

References

1. Shraiman, B.I., Siggia, E.D.: Scalar turbulence. Nature 405(6787), 639–646 (2000)
2. Skvortsov, A., Ristic, B., Morelande, M.: Networks of chemical sensors: a simple mathematical model for optimisation study. In: Fifth International Conference on Intelligent Sensors, Sensor Networks and Information Processing, Melbourne, Australia (2009)
3. Karunasekera, S., Skvortsov, A., Gunatilaka, A.: Minimizing the operational cost of chemical sensor networks. In: Proceedings of the 6th International Conference on Intelligent Sensors, Sensor Networks and Information Processing (ISSNIP), pp. 37–42. IEEE Press (December 2010)
4. Mendis, C., Gunatilaka, A., Skvortsov, A., Karunasekera, S.: The effect of correlation of chemical tracers on chemical sensor network performance. In: Sixth International Conference on Intelligent Sensors, Sensor Networks and Information Processing, Brisbane, Australia (2010)
5. Zhao, F., Guibas, L.: Wireless Sensor Networks: An Information Processing Approach. Morgan Kaufmann Publishers Inc., San Francisco (2004)
6. Demers, A., Greene, D., Hauser, C., Irish, W., Larson, J., Shenker, S., Sturgis, H., Swinehart, D., Terry, D.: Epidemic algorithms for replicated database maintenance. In: Proceedings of the Sixth Annual ACM Symposium on Principles of Distributed Computing, PODC 1987, pp. 1–12. ACM, New York (1987), http://doi.acm.org/10.1145/41840.41841
7. Eugster, P.T., Guerraoui, R., Handurukande, S.B., Kouznetsov, P., Kermarrec, A.-M.: Lightweight probabilistic broadcast. ACM Trans. Comput. Syst. 21, 341–374 (2003), http://doi.acm.org/10.1145/945506.945507
8. Rao, I., Harwood, A., Karunasekera, S.: An unstructured peer-to-peer approach to aggregate node selection. In: Proceedings of the 17th International Symposium on High Performance Distributed Computing, HPDC 2008, pp. 209–212. ACM, New York (2008), http://doi.acm.org/10.1145/1383422.1383450
9. Haas, Z., Halpern, J., Li, L.: Gossip-based ad hoc routing. IEEE-ACM Transactions on Networking 14(3), 479–491 (2006)
10. Gamm, G.U., Sippe, M., Kostic, M., Reindl, L.M.: Low power wake-up receiver for wireless sensor nodes. In: Sixth International Conference on Intelligent Sensors, Sensor Networks and Information Processing, Brisbane, Australia (2010)
11. Gunatilaka, A., Ristic, B., Skvortsov, A., Morelande, M.: Parameter estimation of a continuous chemical plume source. In: Proc. Fusion 2008: 11th International Conference on Information Fusion, Cologne, Germany (2008)
12. Murray, J.D.: Mathematical Biology, vol. 1. Springer, USA (2002)

Analog Molecular Communication in Nanonetworks

Ali Akkaya[1] and Tuna Tugcu[2]

[1] Bogazici University
Department of Computer Engineering
34342 Bebek, Istanbul, Turkey
ali.akkaya@boun.edu.tr
[2] Bogazici University
Department of Computer Engineering
34342 Bebek, Istanbul, Turkey
tugcu@boun.edu.tr

Abstract. Nanotechnology is currently being applied to vast number of fields to overcome the challenges faced with existing technologies, which can not efficiently scale down to nano level. Communication is one of the important problems to be addressed in nano scale environment, and molecular communication is a candidate to address this problem. Existing research on molecular communication concentrates on application of existing digital communication paradigm. In this paper, we approach the problem from another perspective, and propose an analog communication model in which the data is not quantified. The proposed model enables achieving higher data rates using less energy while keeping the error rate bounded. With this characteristics, the proposed method finds promising application options for specific set of communication requirements.

Keywords: analog communication, nanonetworks.

1 Introduction

One of the focuses of Nanotechnology is to construct functional units at nano scale which can perform simple tasks. These functional units are generally referred as nanomachines, and interact with each other to accomplish more complex tasks. For development of these functional units, three main approaches are followed, top-down, bottom-up and hybrid [1].

- **Top-down:** In the top down approach, the aim is to scale down current electronic elements, which are at micro scale, to nano scale. There are several challenges faced in this approach since the physical laws governing nano scale are not the same with micro scale [2].
- **Bottom-up:** In this approach, new nanomachines are constructed from molecular components through chemical molecular reactions.

E. Hart et al. (Eds.): BIONETICS 2011, LNICST 103, pp. 171–182, 2012.

- **Hybrid:** In the hybrid approach, biological components are used as building blocks for nanomachines. Existing biological components are altered and combined to develop more complex systems.

For nanomachines to perform complex tasks, they need to communicate with the external systems and among themselves. The communication network constructed for this purpose is referred as a nanonetwork. Molecular communication is used by many living organisms to enable biological components to communicate with each other, and is one of the methods that can be used for inter-nanomachine communication.

There has been many research efforts about molecular communication in the literature. In [3], molecular communication is introduced and research challenges are explored. One of the listed challenges is creating algorithms for encoding and decoding of information in a robust manner. In our work, we propose an alternative approach to target this challenging issue. In [4], Akyildiz et al. first define the nanomachine and its architecture, and provide a survey on different methodologies on nanonetwork communication for both short range and long range communications. They also list possible applications of nanonetworks. In [5], a model of molecular communication based on gap junction channels is introduced together with a mathematical model. In the model, calcium signaling is used to encode information, which is then transferred from one cell to another via gap junctions. The usage of the model to solve current network design problems such as filtering and switching are also provided in detail. In [6] and [7], two important characteristics of a communication model, channel capacity and noise effect, are investigated for molecular communication. In [8], energy model of a molecular communication model is developed to show how channel capacity can be optimized with the energy limitations.

In all these research work, the model is based on encoding, transmission, propagation, reception and decoding, and aim is to transmit binary information in encoded format. In this paper, we introduce another approach in which data to be transmitted is not quantified, but instead directly sent via molecular diffusion. This approach will enable higher data rates for specific set of applications which can tolerate bounded error. It should be noted that in this paper, we use term "analog" to describe the fact that, similar to analog communication, the value to be transmitted is not digitalized, but the actual number of molecules transmitted is not continuous but discrete.The remainder of the paper is organized as follows, Section 2 gives background information on a generic communication model and details the digital approach, using binary encoding and the proposed communication model. Section 3 discusses the energy consumption and data rate for proposed model. Section 4 provides details on simulation setup we used to analyze the model and the results obtained and we give concluding remarks in Section 5.

2   Molecular Communication Model

In a generic molecular communication model, the aim is to encode information, and transmit it to a propagation medium by a transmitter nanomachine, which is then received by a receiving nanomachine and decoded to access the information. This approach is illustrated in Figure 1.

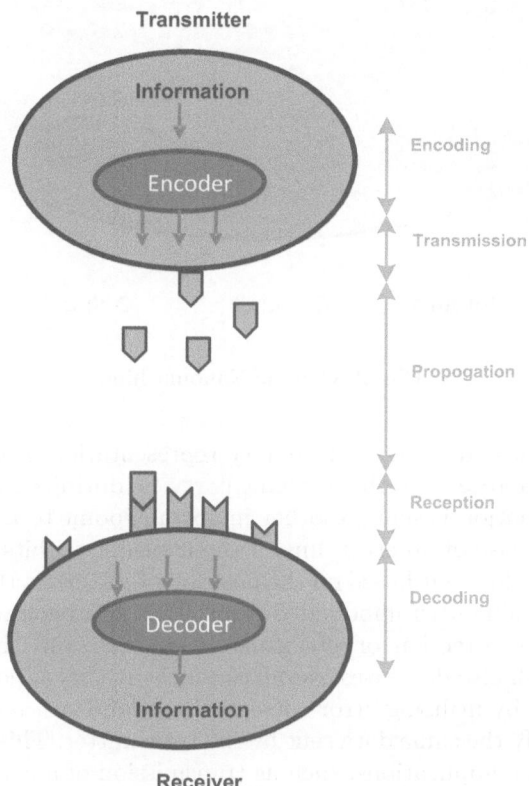

Fig. 1. Communication Model

A nanomachine is a nano scale computational unit that reacts to environmental input to accomplish a task. It takes several input from its environment, and after decoding the inputs, it generates output values. A nanomachine can receive any number of input values, and based on the task it implements, it generates output values. A generic nanomachine according to this definition is shown in Figure 2.

2.1   Digital Molecular Communication Model

In a digital molecular communication model, binary encoding is used. This model can be seen as direct implementation of current digital communication model

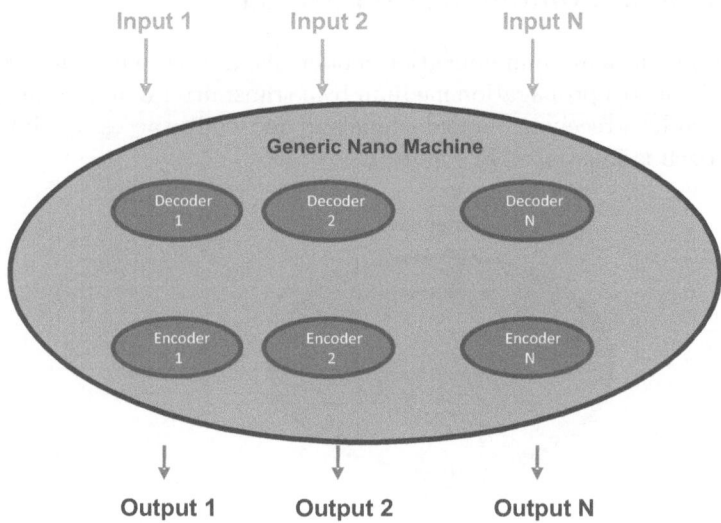

Fig. 2. Generic Nanomachine

to nano scale. With this approach, binary representation of the information is delivered as ones and zeros to the receiving party. So during each symbol duration one binary information is sent. Another important point to note is the effect of channel error. In case of an error, any of the transmitted bits can be corrupted with equal probability, but based on the position of the error, the actual data can be decoded to a value with unbounded error. This approach requires additional error checking and correction, or retransmission to make sure that correct value is transmitted, which also decreases overall channel capacity. The advantage of this approach is that, by utilizing error correction mechanisms, receiver can be sure that it gets exactly the same data sent by the transmitter. This requirement may be crucial for some applications, such as transmission of a command on a DNA code to be executed. However, there may be cases where the exact transmission of the data may not be as important as the data rate achieved.

2.2 Analog Molecular Communication Model

In the analog communication model, each nanomachine sends information not via binary encoding. Instead, during each symbol duration the nanomachine sends specific number of molecules that correspond to the information to be sent. The same structure and mechanisms are used as in the digital model, but the value is directly sent with corresponding concentration level instead of binary encoding. Following steps are repeated at each symbol duration.

- **Encoding:** Before each symbol duration a vesicle that contains same number of molecules, scaled according to the value of the data, is prepared.
- **Transmission:** These molecules are pumped to the propagation medium at start of the symbol duration.
- **Propagation:** The molecules diffuse through the propagation medium. Since it is not a guided transmission medium, the molecules follow the physical characteristics of the medium for propagation.
- **Reception:** Throughout the symbol duration, the molecules that reach the receptors of the receiving nanomachine are successfully received. Each nanomachine has specific receptors, so that only certain molecules are accepted into the nanomachine.
- **Decoding:** The most important part of the cycle is the decoding step. Due to the nature of the propagation medium,
  some molecules do not reach the receiver during the symbol duration. Furthermore, some of the received molecules will be residuals from the previous symbol durations (corresponding to inter-symbol interference), and some others will be generated by other sources (corresponding to noise and interference). Therefore the actual value sent needs to be estimated. The estimation process should take these effects into account. In [10], a statistical model for transmission based on Brownian motion is presented. It is shown that for a given set of molecule transmitted, the average time of first arrival is ∞. In same research, for practical applications, it is advised to define a transmission interval, and declare any particle with higher transmission period as lost. We applied a similar approach and simulation results showed that considering the effect of previous symbol duration will be sufficient since the effect of earlier symbol durations is negligible. This conclusion is also consistent with the results obtained in [8]. To estimate the molecules sent we apply algorithm in Figure 3;

```
Let REMAINING = 0
At each symbol duration do
    NUMBER_OF_MOLECULES_RECEIVED = NUMBER_OF_MOLECULES_RECEIVED - REMAINING
    ESTIMATED_NUMBER_OF_MOLECULES_SENT = estimateSent (NUMBER_OF_MOLECULES_RECEIVED)
    REMAINING = estimateRemining (ESTIMATED_NUMBER_OF_MOLECULES_SENT)
end
```

Fig. 3. Estimation Algorithm

We investigate different options, as detailed in Section 4, for the implementation of *estimateSent* and *estimateRemaining*. The results shows that a linear approximation of the form $ax + b$ is sufficient to model the effect of diffusion. The advantage of the analog approach over digital transmission is shorter symbol durations, lower energy consumption, and better noise resilience.

A data item that is composed of X bits when digitized consumes X symbol durations while the same content can be transmitted in one symbol duration and consume less energy, at the expense of some bounded error in transmission. It should also be noted that the analog communication is limited to a subset of possible input types. Given these characteristics, analog transmission is suitable for some set of communication needs, while digital transmission is suitable for others.

3   Energy Consumption and Data Rate

A mathematical model can be used to analyze energy requirements and data rate for the analog and the digital communication. In both cases, we are interested in the energy and time required to transmit n molecules in a time slot. For the digital method, this corresponds to the transmission of one or few bits (depending on modulation), while in the analog model, the whole value is transmitted. We follow the generic energy model described in [8]. We use the same definitions to model the power consumption in each step in the release of a vesicle;

- E_S: Synthesis of the messenger molecules from their building blocks
- E_V: Production of a secretory vesicle
- E_C: Carrying the secretory vesicles to the cell membrane
- E_E: Releasing the molecules via the fusion of the vesicle and the cell membrane

The total energy consumption E_T to release m molecules is given, accordingly, by

$$E_T = mE_S + \lceil \frac{m}{c_v} \rceil (E_V + E_C + E_E) \tag{1}$$

where c_v is the capacity of a vesicle in terms of the messenger molecules.

3.1   Transmission of an Integer Value

The model defined above can be used to investigate the energy consumption and time required to transmit an integer value X from transmitter to receiver using the digital and the analog models. It is assumed that the integer value is encoded as n bits.

Digital Transmission. In digital transmission, X is transmitted in $n \cdot t_s$. During each t_s, either a vesicle with m_{max} molecules is transmitted or no transmission is done, based on the value of the bit for corresponding t_s, as shown in Figure 4.

To find the expected value of energy consumed to send an arbitrary integer value, let us define P_i as the probability of sending a vesicle with m_{max} molecules at t_{s1}. Then, the expected value of the consumed energy, E_{TOTAL}, is

$$E[E_{TOTAL}] = \sum_{i=0}^{n} P_i \cdot E_T \tag{2}$$

For a random input, P_i will be $\frac{1}{2}$ for $i = 1..n$, and substituting E_T, we have

$$E[E_{TOTAL}] = \frac{n}{2}(m_{max}E_S + \lceil\frac{m_{max}}{c_v}\rceil(E_V + E_C + E_E)) \qquad (3)$$

One should note that the energy consumption is proportional to m_{max}. In [8], it is shown that if the distance between the transmitter and the receiver increases, then t_s and m_{max} also need to be increased to achieve high capacity therefore, to maintain the same capacity level, energy consumption also increases as the distance between the nanomachines increases.

Analog Transmission. In the analog transmission, X is transmitted during a single t_s, in which a vesicle with X molecules will be transmitted as shown at Figure 4.

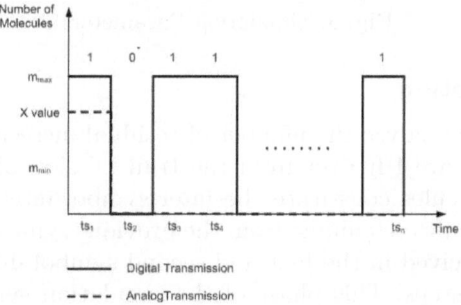

Fig. 4. Molecules sent in digital and analog transmission

To find the expected value of energy consumed to send an arbitrary integer value X, where X is uniformly distributed between 0 and m_{max}, let's define Px_i as the probability of sending value i. Then, $E[X]$ is

$$E[X] = \sum_{i=0}^{m_{max}} Px_i \cdot i \qquad (4)$$

Since X is uniformly distributed, we can write

$$E[X] = \frac{1}{m_{max}+1} \cdot \frac{m_{max}(m_{max}+1)}{2} = \frac{m_{max}}{2} \qquad (5)$$

We can calculate expected value of energy consumption as

$$E[E_{TOTAL}] = \frac{m_{max}}{2}E_S + \lceil\frac{m_{max}}{2c_v}\rceil(E_V + E_C + E_E) \qquad (6)$$

The time required to transmit same data in analog transmission is $\frac{1}{n}$ of the digital method.

4   Simulation and Results

The performance of the proposed method is analyzed by means of the simulator we developed. A transmitter and a receiver nanomachines in a liquid environment are simulated where the molecules propagate according to Brownian motion. The parameters used in the simulation are listed in Figure 5.

Parameter	Value
Temperature	310°K
Viscosity of the fluid	0.001 kg/s.m
Radius of Transmitter	10 μm
Radius of Receiver	10 μm
Radius of Molecule	2.86 nm

Fig. 5. Simulation Parameters

4.1   Data Generation

As the first step, we analyze the number of residual messenger molecules in the environment, which are left over from the transmission of the previous value. These residual molecules constitute the inter-symbol interference. To estimate the number of molecules remaining from the previous symbol duration, the number of molecules received in the first and second symbol durations are recorded to be used in later steps. This phase of the simulation generates the following data;

- Number of molecules transmitted
- Distance between transmitter and receiver
- Symbol duration
- Number of molecules received in the first symbol duration
- Number of molecules received in the second symbol duration

4.2   Estimation Algorithm Selection

With the data set generated, we investigate different algorithms for estimation, starting from simple first order polynomial functions to neural networks and the best-fit curves to the data set [9]. The evaluation is done based on Root Means Square Error (RMSE) of the estimation with respect to the original value transmitted. The results are shown in Figure 6. The advantage of neural networks is the option of having same parameters for all distance-symbol duration pairs, whereas for other algorithms, one needs to fine tune the coefficients for different pairs. When the results are compared, first degree polynomial curve seems to be the most promising one due to its simplicity of implementation, and comparable results to the alternative more complex algorithms. This result implies that the number of molecules transmitted can be estimated with a bounded error using a simple linear polynomial function.

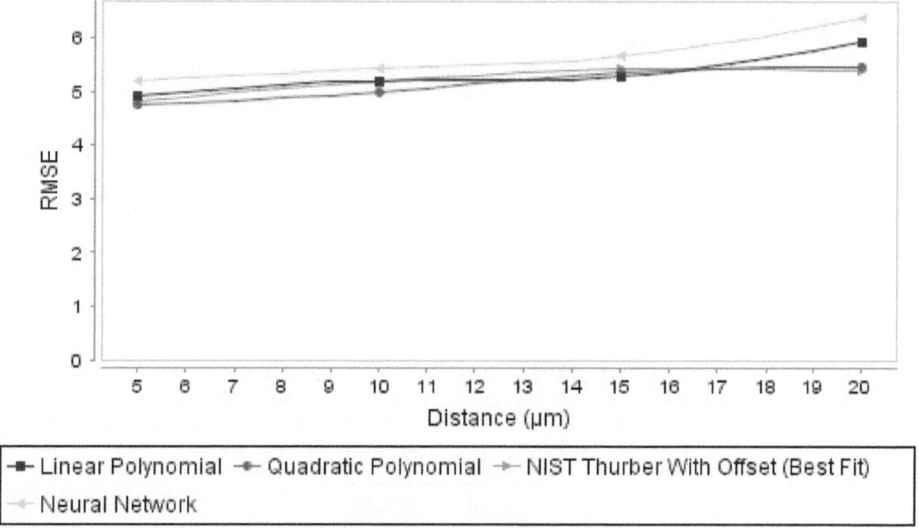

Fig. 6. Estimation Algorithm Selection

4.3   Analog and Digital Communication

We simulate the transmission of a chunk of information from one nanomachine
to another using both the digital and the analog communication methods subject
to the distance and symbol durations. Appropriate symbol duration values for
communication over different distances are taken from [8], which are calculated
using average hitting times of the molecules in simulation.

For the sake of simplicity and generality, we assume that this information
chunk is one byte. In the digital communication method, the value of the byte
value is sent as a sequence of eight bits during eight symbol durations, where
binary "one" is represented with 128 molecules. On the contrary, in the analog
communication method the number of molecules is equal to the byte value sent,
which can be at most 255. With these parameters, the analog communication
method can send the same information eight times faster, using four times less
energy. For better comparison, we analyze the results where the symbol durations and the energy levels are taken to be equal for both methods, and symbol
durations increases in parallel to distance both in digital and analog version.
The results are depicted in Figure 7. For small symbol durations, the digital
communication method is prone to errors, whereas the analog communication
method operates with much smaller error rates. As the distance increases, since
symbol duration increases dramatically, the performance of the digital communication also increases. Note that after 16 micrometers, it is not possible to
receive adequate number of molecules at the receiver; hence communication is
not feasible.

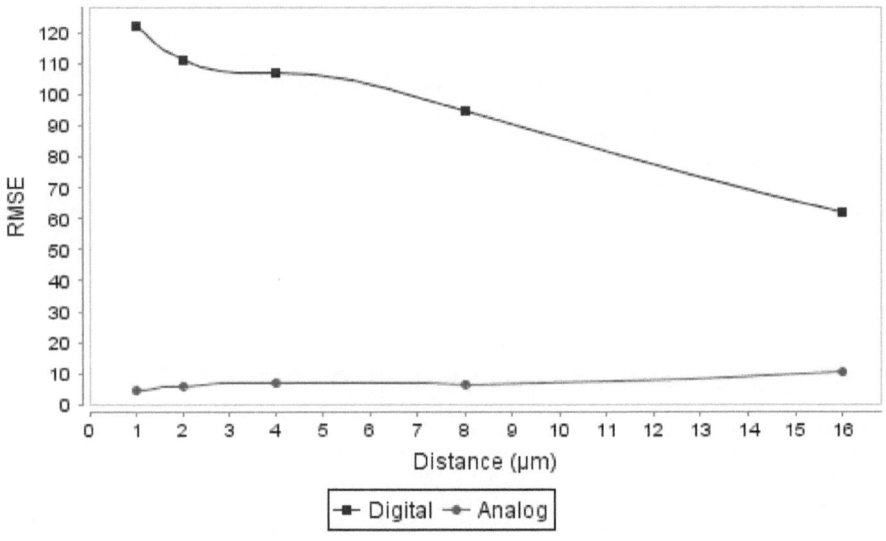

Fig. 7. Analog vs Digital

4.4 Communication Channel with Noise

To show the behavior of the proposed method in communication channel with noise, we consider White Gaussian Noise where molecules are inserted to or removed from the medium. For this purpose, at each time interval a random variable with zero mean and corresponding variance for different noise levels is generated. This value can be a negative or positive. Based on the value of the random variable, molecules are removed from or inserted to the simulation environment. The results for the digital and the analog communication are shown in Figure 8. Both methods experience similar effects under noise, and the error rate increases at a comparable rate.

It is apparent from Figures 7 and 8 that the digital communication does not emerge as a feasible alternative at such noise levels and symbol durations. Therefore, we delve the operational range of the digital communication. Figure 9 shows the results where the analog and the digital communication use the same energy level, i.e., the same number of bits are transmitted in both cases. However, compared to the analog communication, it takes eight times longer for the digital communications to send one byte of information. When there is no noise, the digital communication can send the data without any error, and the analog communication experiences errors due to its communication mechanism. As noise level increases, the digital method starts to experience high error rates, whereas the analog method is more resilient to noise. Thus, the analog method outperforms the digital method if the data needs to be sent in a shorter time or if the noise in the channel is expected to be high.

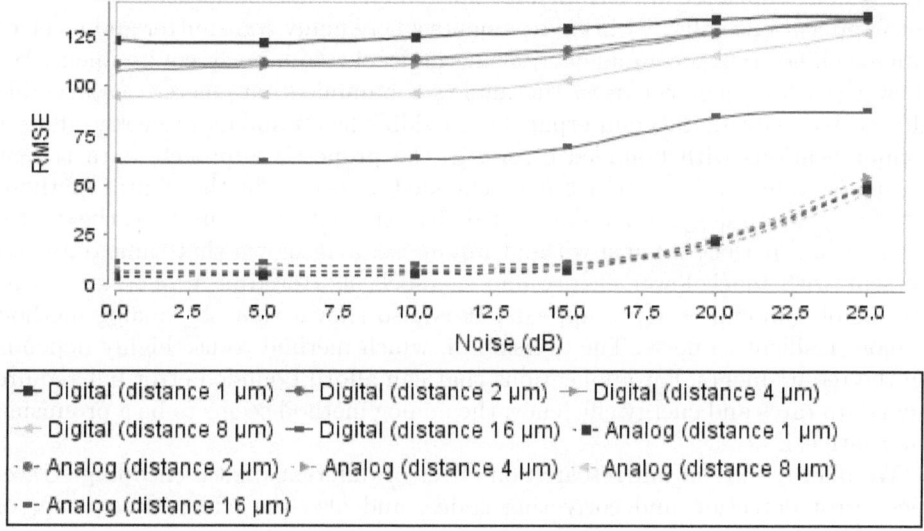

Fig. 8. Noise Effect with Same Durations

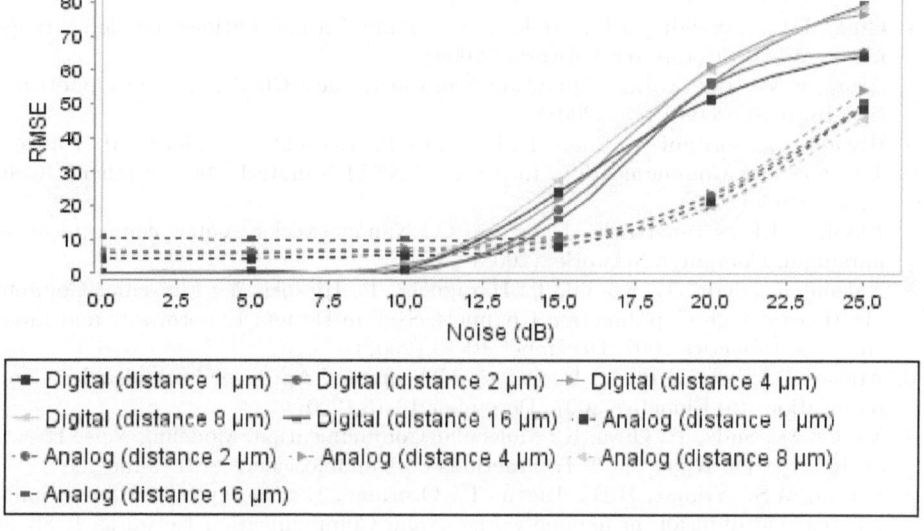

Fig. 9. Noise Effect with Different Durations

5   Conclusion

In this paper, we consider the analog communication as an alternative to the digital communication for specific applications where higher data rate and lower energy consumption can be traded for limited errors in transmission. We believe such applications will exist in environments where many transmitter and receiver nanomachines try to communicate using molecules in a noisy environment. We show that, for such scenarios, the analog communication emerges as a viable alternative to its digital counterpart by providing faster and more energy efficient communications with bounded errors. In the proposed approach, data is sent as analog values rahter than binary encoded values as in the digital method. Although the analog method does not deliver the data as accurate as the digital method in a perfect channel without any noise, it is shown that same data can be sent with much lower energy and in much shorter time frame as soon as reasonable amount of noise appears. It is also shown that the analog method is more resilient to noise. The decision on which method to use highly depends on the requirements. For applications that can afford bounded error but require high data rates and energy efficiency, the analog method seems to be a promising alternative.

We plan to extend our research on effect of different digital encoding strategies, error detection and correcting codes, and also the effect of sequence of transmitted values.

References

1. Gine, L.P., Akyildiz, I.F.: Molecular communication options for long range nanonetworks. Computer Networks (2009)
2. Arora, K.V., Tan, M.L.P.: Quantum Nanoelectronics: Challenges and Opportunities. In: ICSE 2008 Proc. (2008)
3. Hiyama, S., Moritani, Y., Suda, T., Egashira, R., Enomoto, A., Moore, M., Nakano, T.: Molecular Communication. In: Proc. of NSTI Nanotech 2005, Anaheim, California, USA (2005)
4. Akyildiz, I.F., Brunetti, F., Blazquez, C.: Nanonetworks: A new communication paradigm. Computer Networks (2008)
5. Nakano, T., Suda, T., Koujin, T., Haraguchi, T., Hiraoka, Y.: Molecular Communication through Gap Junction Channels: System Design, Experiments and Modeling. In: Bionetics 2007, December 10-13 (2007)
6. Atakan, B., Akan, O.B.: An Information Theoretical Approach for Molecular Communication. In: Bionetics 2007, December 10-13 (2007)
7. Moore, M., Suda, T., Oiwa, K.: Molecular Communication: Modeling Noise Effects on Information Rate. IEEE Transactions on Nanobioscience 8(2) (June 2009)
8. Kuran, M.S., Yilmaz, H.B., Tugcu, T., Ozerman, B.: Energy model for communication via diffusion in nanonetworks. Nano Communication Networks 1, 86–95 (2010)
9. www.zunzun.com (last accessed, April 17, 2011)
10. Eckford, A.: Nanoscale Communication with Brownian Motion. In: Proc. Conf. on Information Sciences and Systems, Baltimore, USA (2007)

Synchronization
of Inhibitory Molecular Spike Oscillators

Michael John Moore and Tadashi Nakano

Graduate School of Engineering, Osaka University,
2-1 Yamadaoka, Suita, Osaka 565-0871, Japan
{mikemo,tnakano}@wakate.frc.eng.osaka-u.ac.jp

Abstract. Molecular communication is the process of transmitting information by modulating the concentration of molecules over time. Molecular communication is suitable for autonomous nanomachines which are limited in size and capability and for interfacing with biological systems which perform functions controlled or influenced by molecules. Some functions may require nanomachines to perform sequential processes. Molecular communication can be used to synchronize multiple nanomachines and to coordinate the timing of the functionality. In this paper, transmitters self-oscillate by releasing a spike of negative autoregulating molecules when concentration of the molecule is below a threshold. When the concentration from a spike disperses and decreases below the threshold, the transmitter releases another spike of molecules. When the environment includes two transmitters, the oscillations of the two transmitters achieve in-phase or anti-phase synchronization depending on the distance between the transmitter and receiver. When there are multiple transmitters arranged in a circle, the oscillations of the transmitters produce in-phase or partially in-phase synchronization. Simulations were performed to characterize the period of oscillation and the phase difference in the oscillations of multiple transmitters.

Keywords: Molecular Communication, Spike Signal, Oscillation, Synchronization, Negative Autoregulation.

1 Introduction

Molecular communication is the process of transmitting information by modulating the concentration of molecules (e.g. calcium ions, peptides) over time [1,2]. Molecular communication is suitable for nanomachines which are limited in size and capability and for interfacing with biological systems which perform functions controlled or influenced by molecules. A nano- to micro-scale biological transmitter is, for example, a protein complex, a bacterium, or a cell. Examples of applications which can be influenced through molecular communication include medical applications for interacting with cells, environmental applications for detecting and processing waste molecules, and manufacturing applications for rearranging molecules using self-organizing processes [2,3].

E. Hart et al. (Eds.): BIONETICS 2011, LNICST 103, pp. 183–195, 2012.

To perform molecular communication, (1) a transmitter nanomachine, T, encodes information into molecules, (2) T releases the molecules, (3) the molecules propagate through an aqueous environment, (4) a receiver nanomachine, R, reacts to the molecules, and (5) finally R decodes information from the molecules.

This paper focuses on using molecular communication to synchronize oscillating nanomachines. A nanomachine periodically increases and/or decreases the concentration of a molecule, and causes the concentration of the molecule to oscillate over time. The oscillator model in this paper assumes molecules are released at a single time instant to produce a spike of concentration increase. For example, a cell releases the contents of a vesicle resulting in the release of many molecules at the same time [7]. The molecules are negative autoregulating. A negative autoregulating molecule type A is a molecule which inhibits the production of itself (i.e., other molecules of A). While the production of A is inhibited, T does not spike again. The concentration of molecules decreases as the result of diffusion into the environment. When the concentration of molecules decreases below a threshold, T releases the next spike of molecules.

The oscillator model considered in this paper is relatively simple in order to focus on analyzing how diffusion processes can synchronize multiple transmitters. We also identify conditions to produce in-phase and anti-phase synchronization in the case of two oscillators and a small number of oscillators arranged in a circle. Two oscillators in anti-phase synchronization activate at the same frequency, but alternate activation half-way out of phase.

Section 1.1 describes related work on synchronization. Section 2 describes general parameters of molecular communication which may impact the synchronization and section 3 describes simulations of the protocols to evaluate the oscillation and synchronization.

1.1   Synchronization by Molecule Diffusion

In biology, oscillation often occurs as the result of positive or negative feedback loops. Examples of biological oscillations include calcium concentration in cells, inter-cellular communication, and circadian rhythms [4]. For example in calcium signaling [4], cells oscillate the concentration of calcium. The release of calcium in one cell can cause the release of calcium in a nearby cell and results in synchronized release of calcium from the cells. Another example of synchronization is neurons coupled through delayed inhibitory feedback [6]. This type of oscillation can be anti-phase depending on the timing the inhibitory feedback process.

In systems biology, an oscillator motif is composed of two molecules which produce a delayed negative feedback loop [5]. A negative feedback loop is expected to be a simple way to model oscillation since it can be produced by using only few types of molecules [5]. Also, oscillators based on inhibitory reactions can exactly synchronize [6]. This is unlike excitatory oscillators which spike at times offset by propagation delay.

Recent research on synchronization includes observation and design of nanomachines coupled through diffusion. [8] describes oscillation observed in an array of micro-oscillators which perform BZ chemical reactions. The micro-oscillators are coupled primarily through inhibitory reactions and produced a variety of synchronization regimes including in-phase and anti-phase synchronization and stationary Turing patterns. Different synchronization behaviors may be useful for different applications of molecular communication. [9] proposes modifying bacterial quorum sensing to synchronize nanomachines to the same state, to perform some operation, and then repeats the synchronization.

2 Transmission Model

General parameters of T (i.e., a transmitter nanomachine), R (i.e. a receiver nanomachine), and the environment impact the ability for T and R to achieve synchronization. For simplicity, T and R are assumed to not move relative to the environment.

2.1 Spike Transmission

T can communicate to R by releasing a predetermined type of molecule at some rate over time. Nanomachines can generate simple signal patterns such as releasing many molecules in a spike (e.g. T is a cell and releases the contents of a vesicle so that many molecules are released into the environment over a short time interval [7]), releasing molecules at a constant rate (e.g. T is an enzyme at saturation and catalyzes a chemical reaction to produce molecules over time), or releasing molecules at some frequency of regularly oscillating concentration (e.g. T is a cell and produces waves of high and low calcium concentration [10]). This paper focuses on synchronizing T and R using a spike signal. A spike signal is expected to perform as the ideal case for diffusion since molecules are all released at a single time instant, and thus variance in arrival times between the molecules in the spike is minimized [11].

The type of molecules released, e.g. molecules of type A, is assumed to be negative autoregulating. Thus after transmitting a spike of A, T enters a refractory period during which T does not transmit A. T can use the refractory period to prepare for another transmission or to avoid transmitting multiple spikes in response to a single signal. Also, if T waits for prior concentration of A to become sufficiently low, then noise from prior concentration of A will be reduced.

Fig. 1 illustrates a chemical network for producing a refractory period with A. Molecules of type a are in the transmitter at a constant concentration. Chemical reaction of a results in the production of a single spike of type A molecules. A inhibits the production of A through negative autoregulation. Thus, molecules of type a should not produce another spike of A until A decreases in concentration. Note that the quick release of many molecules as a spike (e.g., as the result of releasing molecules from a vesicle) is not described by this chemical network.

2.2 Signal after Diffusion

Diffusion is a common model of propagation for molecular communication. Thermal noise in the environment (e.g. molecules in the environment are in the

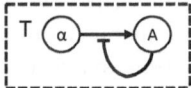

Fig. 1. T contains molecules of type a which produce a single spike with molecules of type A. Type A molecules negatively autoregulate to produce a refractory period.

liquid or gas phase) mixes molecules and causes molecules to diffuse through-out the environment [12]. To simplify analysis in this paper, molecules propa-gate by diffusion in the environment and other propagation, such as fluid flows, are assumed to be negligible. The environment can include molecular motors [13] to guide molecules or repeater nanomachines [14] which detect and amplify molecule concentrations; however, for simplicity of protocol analysis, this paper focuses on propagation by diffusion without additional components for enhancing propagation.

Assume a spike of N molecules is released at location in an unbounded 3-dimensional space. $C(x,t)$ describes the concentration distribution of molecules at distance x from the location of release and at time t. $C(x,t)$ follows a Gaussian distribution with zero mean and a standard deviation of $\sigma = \sqrt{2Dt}$ where D is the diffusion coefficient of the type of molecules [12].

$$C(x,t) = \frac{N}{(4\pi Dt)^{3/2}} e^{-\frac{x^2}{4Dt}}. \tag{1}$$

If T is an oscillator and spikes at regular intervals, then the average molecule release is similar to releasing molecules at a constant rate. [16] describes a steady-state case in which T releases molecules at a constant rate of molecules per unit time over the time range $[0, \infty]$ and concentration at the steady state can be calculated as a finite value. If the steady-state concentration at T can be approximated as some finite value, there should exist some frequency which allows T to decrease below a finite concentration threshold.

2.3   Receiving a Spike Signal

A signal after propagation produces a different signal pattern at a receiver po-sitioned at a distance x following Eq.(1). In this paper for simplicity, T and R react when the concentration of molecules passes a constant threshold, H. For example, R has a chemical process which is likely to activate when the con-centration of molecules at R is below a concentration threshold. A reasonable threshold is required since the concentration detected at R varies over time.

In this paper, T and R communicate using negative autoregulating molecules. Fig. 2 illustrates T and R using the same type of negative autoregulating molecule. If T releases a spike of molecules, T can prevent R from activating by increasing the concentration of negative autoregulating molecules at R, and vice versa.

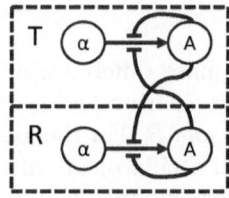

Fig. 2. Oscillators T and R coupled by using the same type of molecule

3   Numerical Simulation

3.1   Simulation Model and Configuration

The simulation environment is an infinite 3-dimensional space. 3-dimensions was chosen since oscillation frequency λ is expected to be stable in a 3-dimensional space. T and R are each modeled as single points in space. A single spike is assumed to follow an impulse described by a Dirac delta function at time t. This model does not consider the noise in timing of molecule release from a vesicle; however, the time necessary to release molecules from a vesicle is typically much faster relative to the time necessary for diffusion-based propagation. Single spike transmissions are recorded in an array of tuples (x, y, z, t). Each tuple represents parameters for the concentration equation Eq. (1). Parameters N and D in the equation are assumed to be constant for all spikes. For example, if T spikes at time t_1, then an entry (x_T, y_T, z_T, t_1) is added to the array where (x_T, y_T, z_T) are the coordinates of T. Note that the simulation running time for this technique is dependent on the number of spikes transmitted and not on the number of molecules released in each spike.

The array of equations can be used to calculate the concentration at any receiver R at a later time.

$$C^*(R, t) = [\sum_{i=0}^{j} C(x_i, t - t_i)] \tag{2}$$

where j is the number of entries in the array, x_i is the Euclidean distance from the receiver R to the source of the ith spike of molecules, t is greater than all times t_i recorded in the array. Concentration at T and R are assumed to be the concentration at their respective points in space.

The time step of the simulation is 0.01 sec. The simulated time was chosen as 3600 sec which was a sufficient amount of time to observe stability for two oscillators. The diffusion coefficients for all types of molecules were chosen to be 10 $\mu m^2/second$ which is the diffusion coefficient of a protein in water. Unless stated otherwise, N was arbitrarily chosen as 8192 which is a feasible number of molecules to release from a vesicle to produce a spike. Unless stated otherwise, T and R activate when concentration is below threshold $H = 0.1$ $molecules/\mu m^3$. H corresponds to an arbitrarily chosen low concentration of molecules. Data points are averaged over 100 simulation runs.

3.2   Simulation Metrics

In simulation, the period and phase difference are measured to determine how close the system is to an ideal oscillation and synchronization.

Period λ between spikes by T and R: For the ideal oscillator, the time between consecutive spikes, λ, at a given oscillator are always the same amount of time. However, λ for the oscillator, for example T, can vary over time in simulation. Assume that T spikes at times $[t_{T,0}, t_{T,1}, ..., t_{T,n}]$. The period of time between spikes for the jth spike at T is $\lambda_{T,j} = t_{T,j} - t_{T,j-1}$.

Phase difference between T and R: Phase difference describes the difference in phase between when T and R spikes. The ideally synchronized oscillators spike exactly at the same time (i.e., a phase difference of 0.0) for in-phase synchronization or half-way between phase (i.e., a phase difference of 0.5) for anti-phase synchronization. The phase of the jth spike by R is measured relative to the jth spike by T and the period between spikes at T. In simulation, phase difference ϕ_j is measured as

$$\epsilon_j = \frac{(t_{R,j} - t_{T,j})}{\lambda_{T,j}} \bmod 1 \tag{3}$$

The specific order of spikes is assumed to not matter. For example, assume T spikes at times $t = 0$ and $t = \lambda$ and R spikes at time $t = 0.9\lambda$, then phase difference is 0.1. This can be calculated as

$$\phi_j = min(\epsilon_j, 1 - \epsilon_j). \tag{4}$$

Thus, the maximum phase difference measured in simulation is 0.5.

3.3   Single Oscillator

Fig. 3 illustrates the time between spikes, λ, over time for a single transmitter. The transmitter based on negative autoregulating spikes produces oscillating concentration. λ increases over time as concentration from prior spikes disperses more and more slowly. λ also appears to converge to some value which indicates that the model produces a clock-like regular oscillation. Convergence of period λ takes time since the rate of molecule dispersion is changing over time.

3.4   Coupled Oscillators

Fig. 4 illustrates the difference in phase between T and R separated by 20 μm and 40 μm. T and R start with random phase. Random phase was produced by initializing concentration at each oscillator to correspond to a single spike randomly in the range $[0, 20]$ seconds before the start of simulation. This time range corresponds approximately to the time between the first and second spike of a single oscillator. After a short amount of time, T and R synchronize.

Different types of synchronization were observed depending on distance. This occurs since oscillators coupled by inhibitory interactions can produce in-phase

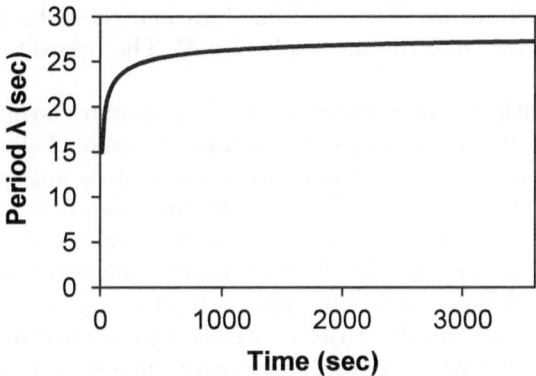

Fig. 3. T oscillates by spiking with negative autoregulating molecules

or anti-phase synchronization depending on the timing of the inhibitory reaction [6]. In this simulation, the timing of inhibitory actions depends on the distance between T and R. Three cases were observed: at 20 μm T and R were in-phase; at 20 μm T and R were anti-phase; and at 40 μm T and R were in-phase. At 40 μm, anti-phase was not observed. At 20 μm, a ratio of 0.24 were in-phase and 0.76 were anti-phase. In the case of 20 μm, if T and R start nearly in-phase, they converged to in-phase.

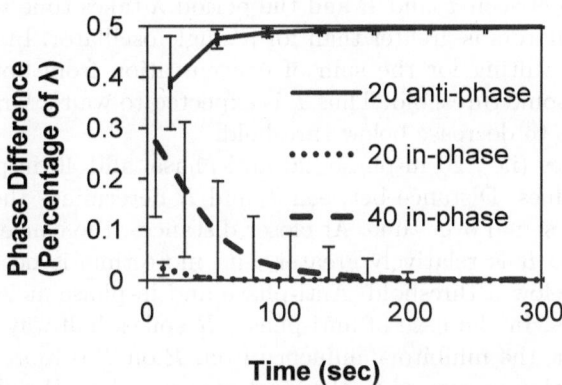

Fig. 4. Phase difference over time for oscillators T and R separated by 20 μm or 40 μm. Anti-phase and in-phase were observed for μm.

In all simulations with 40 μm between T and R, the pair converged to in-phase synchronization. The specific timing of when inhibition arrives and how long inhibition lasts impacts convergence. If T spikes earlier than R, then the next spike by T will be more in-phase. Since T spikes earlier, then the inhibition from T on R will arrive relatively earlier than the inhibition from R on T.

If both inhibitory effects last for a similar duration, then the next spike by T will be relatively later than the next spike by R. The opposite case of T later than R is similar.

In simulations with 20 μm between T and R, anti-phase synchronization can occur when T and R are out of phase. Assume T spikes before R at time t_0 and the spike inhibits R starting from time t_1. If R does not spike before time t_1 then T blocks the activation of R before R can spike. Concentration is then relatively higher at T than R. Then, R spikes next since it decreases below H first. Then the process repeats with T and R performing opposite functions after each spike. Phase difference at 20 μm appears to converge more quickly than at 40 μm, which may be a result of the relatively higher concentration influence from T on R and vice versa. Higher concentration results at closer distances since concentration is inversely proportional to distance in Eq. 1.

At 20 μm between T and R, T and R can converge to in-phase when they start nearly in-phase. When T and R are nearly in phase, then a spike, for example, from T requires some amount of time to propagate to R; however, R has already spiked since it is in nearly the same phase as T and T does not prevent the spike at R. Then the timing of subsequent spikes similarly follows the in-phase case of 40 μm between T and R.

Fig. 5 illustrates λ for T and R separated by 20 μm or 40 μm. Similar to the single oscillator, λ initially increases and then appears to flatten. The time between spikes for the case of two oscillators is greater than a single oscillator. Convergence of phase difference takes time since a sufficient number of molecules must propagate between T and R and the period λ takes time to stabilize.

λ for two oscillators is greater than for a single oscillator. In the case of two oscillators, T is waiting for the sum of concentration from both T and R to decreases below some threshold. Thus T is expected to wait a longer for the sum of concentrations to decrease below threshold.

The three cases (i.e., 20 in-phase, 20 anti-phase, and 40 in-phase) converge to different λ values. Distance between T and R determines the concentration of the inhibitory signal over time. At closer distances, the concentration contribution from T to R is relatively greater, and more time is necessary for their sum to reduce below a threshold. Anti-phase and in-phase at 20 μm also have different λ values. In the case of anti-phase, R spikes half-way between spikes by T. As a result, the inhibitory influence from R on T is more recent and has higher concentration compared to the in-phase case when R spikes at the same time as the previous spike of T. Therefore, T must wait slightly longer for the concentration to decrease below threshold.

Fig. 6 illustrates the standard deviation in period λ over time for the three cases. The standard deviation of λ quickly decreases towards 0 in all cases.

3.5   Circle of k Oscillators

In some applications, it may be necessary to use several oscillators (e.g. to increase the range of synchronization). To evaluate scalability in terms of the number of oscillators, we assume k oscillators are positioned evenly on a circle

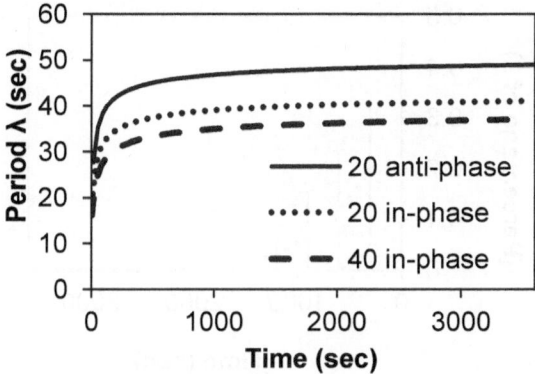

Fig. 5. Phase λ over time for oscillators T and R separated by 20 μm or 40 μm. Anti-phase was not observed at 40 μm.

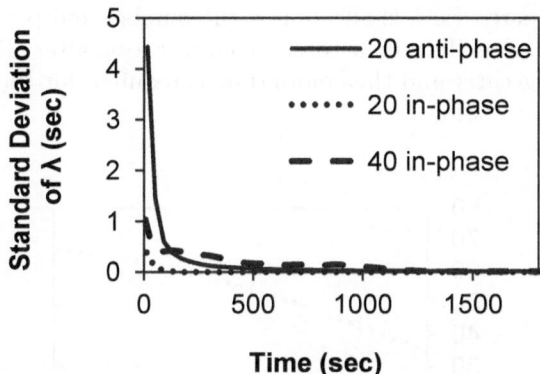

Fig. 6. Standard deviation in phase λ over time for oscillators T and R separated by 20 μm or 40 μm.

with adjacent oscillators separated by distance r. A circle is expected to favor in-phase synchronization since each oscillator has the same distances to other oscillators. The size of the circle increases as k increases such that adjacent oscillators on the circle are always 50 μm apart. 50 μm was chosen in order to avoid anti-phase synchronization between adjacent oscillators.

Fig. 7 illustrates the phase difference over time in the case of $k = 4$. Initially phases are randomized resulting in an average phase difference of 0.25 with a relatively high standard deviation. If oscillators are not synchronized, then the phase difference is expected to be uniformly distributed from 0.0 to 0.5 which results in a mean of 0.25. Over time, the phase difference converges to in-phase.

Fig. 8 illustrates λ for circles with different numbers of oscillators. Time between spikes, λ, and phase difference were averaged during the simulation time range of 3300 to 3600 seconds which is expected to represent a partially converged state.

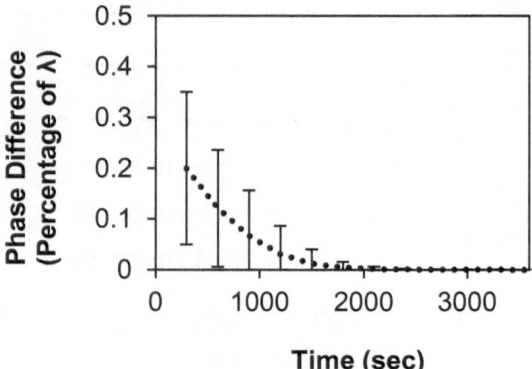

Fig. 7. Mean and standard deviation in phase difference with k=4 oscillators separated by 50 μm

λ increases as k increases. λ is affected by the number and position of oscillators. As the number of oscillators increases, concentration after all the oscillators spike is likely to be greater and thus more time is required for the concentration to decrease below threshold.

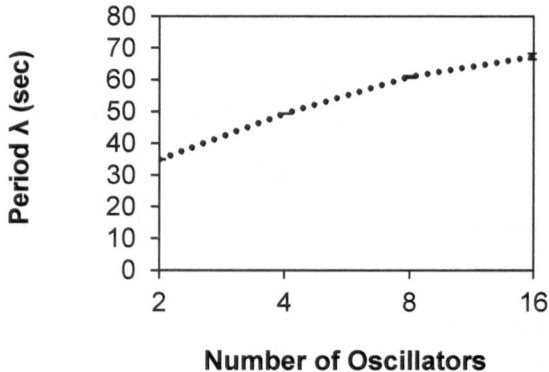

Fig. 8. Mean time between spikes, λ, for oscillators in a circle separated by 50 μm

Fig. 9 illustrates the mean and standard deviation of phase difference for circles with different numbers of oscillators (k = 2, 4, 8, 16). Phase difference of each oscillator is measured relative to one arbitrarily chosen oscillator.

The k oscillators achieve in-phase synchronization for small k and partially in-phase synchronization for larger k. Initially all oscillators start in a random phase with a mean 0.25 phase difference. Then, oscillators gradually oscillated more in-phase in all cases. As k increased, the oscillators at the end of simulation were relatively less in-phase. Since phase is most strongly influenced by adjacent oscillators, corrections in phase timing must propagate around the circle. If corrections propagate over many hops, then oscillators may remain out of phase longer. With very large circles, the phase difference is likely to remain relatively uniformly random (i.e., have a mean 0.25) for a long period of time.

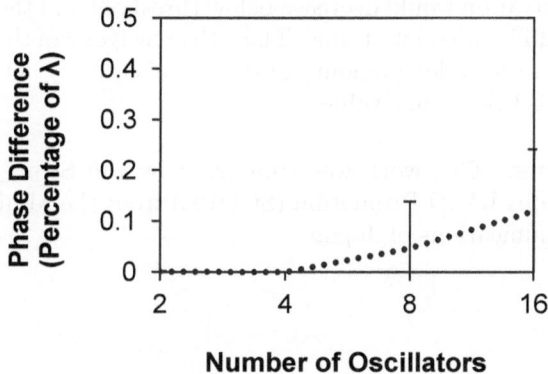

Fig. 9. Mean phase difference between k oscillators in a circle with 50 μm between adjacent oscillators

4   Conclusions

Nanomachines in molecular communication protocols may benefit from oscillating clocks to control the timing of chemical processes. When there are multiple oscillators, it may be beneficial to synchronize the oscillators to coordinate the nanomachine chemical processes.

In this paper, we model an oscillator as a transmitter nanomachine which releases spikes of negative-autoregulating molecules. The oscillation considered in this paper is relatively simple in order to focus on how diffusion impacts synchronization. When there are two of the oscillators, the oscillators produce in-phase or anti-phase synchronization depending on the distance and initial phase difference between the oscillators. When there are a circle of the oscillators spaced sufficiently far apart, the oscillators converge to in-phase or partially in-phase synchronization depending on the number of oscillators in the circle. These results indicate that diffusion of a single type of inhibitory molecule may be sufficient to produce different synchronization characteristics.

Future work includes evaluating the robustness of the oscillation and synchronization to various parameters such as noise, distance, initial conditions of T and R, and transmission characteristics of T and R. For example, concentration may pass the threshold earlier or later than expected as a result of noise, and thus the period and phase of oscillation may vary widely. R can sense the concentration of molecules over some duration to increase the accuracy of the concentration detected [17]. For example, the theoretical uncertainty in sensed concentration by a receiver which perfectly monitors concentration is $3/(5\pi DacT)$ [17] where a is the radius of the spherical monitor, c is the concentration of the molecule, and T is the duration of sensing time. Depending on parameter values, only a few seconds of monitoring may be necessary to achieve an accurate measure. In the case of diffusion, the peak concentration at R arrives at time $r^2/6D$ [18]. After the peak arrives, concentration is strictly decreasing. Assuming that there are no further molecular signals, the concentration would decrease below threshold and then remain below threshold for an infinite amount of time. Thus, the receiver can theoretically monitor the environment for a long amount of time to accurately determine whether the concentration is below some value.

Acknowledgments. This work was supported by the Strategic Information and Communications R& D Promotion (SCOPE) from the Ministry of Internal Affairs and Communications of Japan.

References

1. Hiyama, S., Moritani, Y., Suda, T., Egashira, R., Enomoto, A., Moore, M., Nakano, T.: Molecular Communication. In: Technical Proceedings of the 2005 NSTI Nantechnology Conference and Trade Show, vol. 3, pp. 391–394 (2005)
2. Moore, M., Enomoto, A., Suda, T., Nakano, T., Okaie, Y.: Molecular Communication: New Paradigm for Communication among Nano-scale Biological Machines. In: The Handbook of Computer Networks, vol. 3, pp. 1034–1054. John Wiley and Sons Inc. (2007)
3. Akyildiz, I.F., Brunetti, F., Blazquez, C.: Nanonetworks: A New Communication Paradigm. Computer Networks Journal 52, 2260–2279 (2008)
4. Goldbeter, A.: Computation Approaches to Cellular Rhythms. Nature 420, 238–245 (2002)
5. Alon, U.: An Introduction to Systems Biology Design Principles of Biological Circuits. Chapman & Hall, Boca Raton (2007)
6. Ernst, U., Pawelzik, K., Geisel, T.: Delay-induced Multistable Synchronization of Biological Oscillators. Phys. Rev. E 57, 2150–2162 (1998)
7. Mosharov, E., Sulzer, E.: Analysis of Exocytotic Events Recorded by Amperometry. Nature Method 2(9), 651–658 (2005)
8. Toiya, M., Gonzalez-Ochoa, H.O., Vanag, V.K., Fraden, S., Epstein, I.R.: Synchronization of Chemical Micro-oscillators. The Journal of Physical Chemistry 8(1), 1241–1246 (2010)
9. Abadala, S., Akyildiz, I.F.: Automata Modeling of Quorum Sensing for Nanocommunication Networks. Nano Communication Networks 2(1), 74–83 (2011)
10. Grachevanw, M.E., Gunton, J.D.: Intercellular Communication via Intracellular Calcium Oscillations. J. Theor. Biol. 221, 513–518 (2003)
11. Garralda, N., Llatser, I., Cabellos-Aparicio, A., Pierobon, M.: Simulation-based Evaluation of the Diffusion-based Physical Channel in Molecular Nanonetworks. In: 2011 IEEE Conference on Computer Communications Workshops, pp. 443–448 (2011)
12. Berg, H.C.: Random Walks in Biology, pp. 22–23. Princeton University Press (1993)
13. Moore, M., Suda, T., Oiwa, K.: Molecular Communication: Modeling Noise Effects on Information Rate. IEEE Transactions on Nanobioscience 8(2), 169–180 (2009)

14. Nakano, T., Shuai, J.: Repeater Design and Modeling for Molecular Communication Networks. In: Proceedings IEEE INFOCOM (International Conference on Computer Communications) Workshops, pp. 501–506 (2011)
15. Pierobon, M., Akyildiz, I.F.: Diffusion-based Noise Analysis for Molecular Communication in Nanonetworks. IEEE Transactions on Signal Processing 59(6), 2532–2547 (2011)
16. Bossert, W.H., Wilson, E.O.: The analysis of olfactory communication among animals. Journal of Theoretical Biology 5(3), 443–469 (1963)
17. Endres, R.G., Wingreen, N.S.: Accuracy of Direct Gradient Sensing by Single Cells. Proceedings of the National Academy of Sciences 105(41), 15749–15754 (2008)
18. Moore, M.J., Nakano, T., Enomoto, A., Suda, T.: Measuring Distance with Molecular Communication Feedback Protocols. In: 5th ACM/ICST Int. Conf. Bio-Inspired Models of Network, Information, and Computing Systems (2010)

Propagation Delay of Brownian Molecules in Nano-Biosensor Networks

Yutaka Okaie, Michael John Moore, and Tadashi Nakano

Graduate School of Engineering, Osaka University,
2-1 Yamadaoka, Suita, Osaka 565-0871, Japan
{yokaie,mikemo,tnakano}@wakate.frc.eng.osaka-u.ac.jp

Abstract. In the emerging area of nanonetworks, nano to micro-scale devices called nanomachines are deployed to perform various tasks for applications [1][2]. In this short paper, we consider a biosensor network that consists of nano-scale biosensors; i.e., sensors capable of sensing chemical signals (e.g., toxic chemical substances). In the biosensor network, stationary sensors are distributed over a two dimensional space, and expected to capture a chemical signal that appears in the space and that propagates via Brownian motion. We employ two different placement schemes to distribute sensors, and measure the propagation delay that is required to detect a chemical signal. Preliminary simulation results are provided to show the impact of placement schemes as well as the number of sensors on the propagation delay.

Keywords: nanonetworks, nano-biosensors, placement schemes.

1 Overview

In this paper, we consider a nano-biosensor network that consists of a number of nanomachines with sensing capabilities. Nanomachines, simply referred to as sensors in this paper, are deployed in the aqueous environment of interest for monitoring chemical signals. For instance, sensors may be deployed in the human body for monitoring toxic chemical substances, disease-indicating molecules, viruses, or pathogens that potentially appear in the human body. The specific tasks for sensors or a group of sensors are to identify a potential threat and coordinate to remove the threat or report to an outside system via wireless sensor networks.

One of the design goals of nano-biosensor networks is to find sensor configurations to minimize propagation delay of the signal, i.e., the time from when a signal arrives in the area to the time when it's detected by a sensor. We conduct simulation study to understand the impact of the sensor configurations such as the distribution and the number of sensors on propagation delay.

The problems considered in this paper are related to sensor placement problems for wireless sensor networks (WSNs). A variety of strategies are designed for WSNs to improve various measures [3]. Our work differs from the existing studies for traditional WSNs in that we consider nanonetworks. The target signal

E. Hart et al. (Eds.): BIONETICS 2011, LNICST 103, pp. 196–198, 2012.

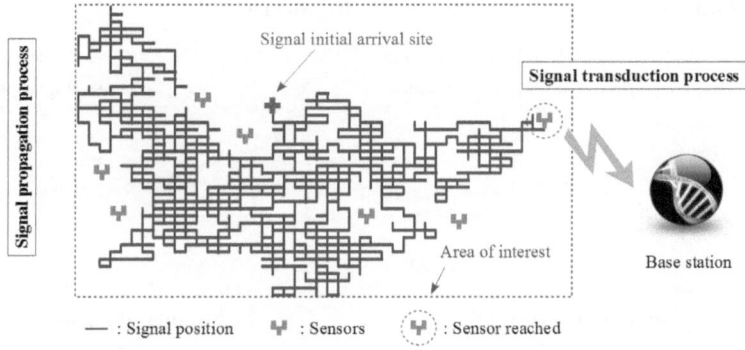

Fig. 1. Overview of the nano biosensor networks

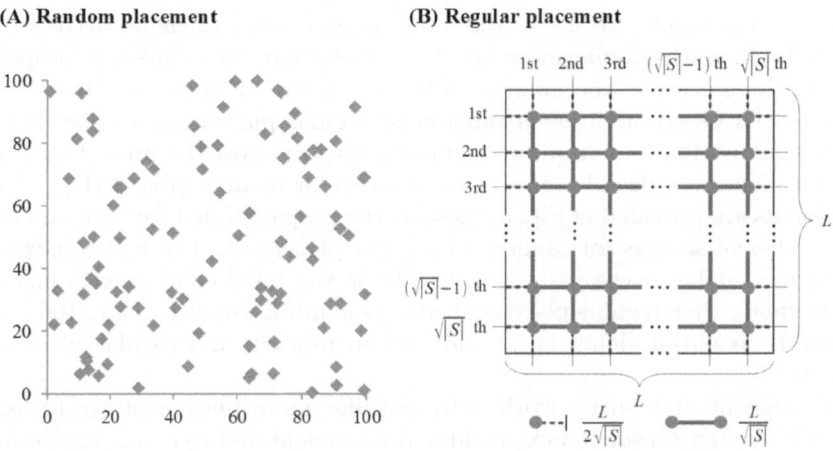

Fig. 2. Random and regular placement

in a nanonetwork consists of molecules in an aqueous environment, the fundamental behavior follows Brownian motion, and the signal is detected via physical binding to a sensor.

2   Preliminary Results

Fig. 1 shows the model of the nano-biosensor network. The model includes the area of interest, signal, stationary sensors, and the base station. The area is the environment where signals arrive and propagate via Brownian motion. Stationary sensors, simply called sensors, are nano-scale biosensors made of biological materials such as receptors and enzymes. Sensors are capable of detecting specific chemical signals and statically deployed in the area. Sensors detect chemical signals when they physically contact the chemical signals, and notify the base station of the signal arrival via external signaling pathways.

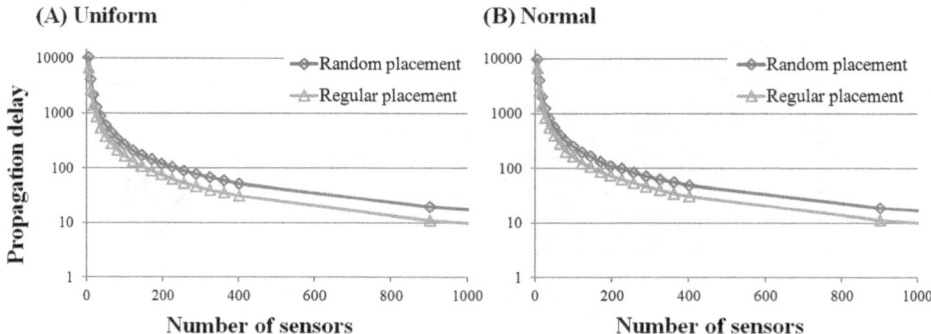

Fig. 3. Propagation delay when signal arrival follows (A) uniform and (B) normal distributions

In the simulation, we use a 100×100 square area. A signal arrives in the area following either uniform or normal distribution, and randomly propagates until it hits a sensor. The number of sensors is varied from 1 to 1000. Sensors are deployed based on either a random or regular placement scheme. Random placement distributes sensors in a uniform manner over the area (Fig. 2 (A)). Regular placement distributes sensors in a regular manner (Fig. 2 (B)). Simulation results are provided in Fig. 3, showing the propagation delay as a function of the number of sensors for random or regular placement. The figure shows that propagation delay decreases exponentially as the number of sensors increases. It also shows that regular placement always exhibits smaller delay; 196 sensors achieve propagation delay 112.2 and 74.2 in random and regular placements, respectively.

Our current and future work is to consider more placement strategies, robustness against sensor failure, a three-dimensional and dynamic environment, distributed and self-organizing algorithms for mobile sensors, and target signals that can actively migrate in the environment.

Acknowledgement. This work was supported by Grant-in-Aid for Young Scientists A (No. 22680006) from the Japan Society for the Promotion of Science (JSPS) and by the Strategic Information and Communications R&D Promotion (SCOPE) from the Ministry of Internal Affairs and Communications of Japan.

References

1. Hiyama, S., Moritani, Y.: Molecular communication: Harnessing biochemical materials to engineer biomimetic communication systems. Nano Communication Networks 1, 20–30 (2010)
2. Akyildiz, I., Brunetti, F., Blazquez, C.: Nanonetworks: a new communication paradigm. Computer Networks 52, 2260–2279 (2008)
3. Younis, M., Akkaya, K.: Strategies and techniques for node placement in wireless sensor networks: A survey. Ad Hoc Networks 6, 621–655 (2008)

Co-Channel Interference for Communication via Diffusion System in Molecular Communication

Mehmet Şükrü Kuran and Tuna Tugcu

Bogazici University,
Department of Computer Engineering,
34342 Bebek, Istanbul, Turkey
{sukru.kuran,tugcu}@boun.edu.tr

Abstract. In this paper, we show the detrimental effects of Co-Channel Interference (CCI) in Molecular Communications in the context of the Communication via Diffusion (CvD) system. The effects of CCI are evaluated with respect to system performance parameters, probability of hitting to the intended receiver and the channel capacity, while considering additional environmental affects such as the Inter symbol Interference (ISI). Based on our simulation results in a 3D diffusion environment, we conclude that similar to classical wireless communication systems, CCI is an important source that adversely affects the performance in a CvD system. Also, we show that a molecular reuse range concept which is analogous to the frequency reuse range in wireless communications, can be used to cope up and control the severity of CCI where necessary.

Keywords: nanonetworks, molecular communication, communication via diffusion, interference.

1 Introduction

Molecular communication (MC) is an emerging communications paradigm that aims to provide solutions for communication between nano- or micro-scale devices, or in another words, the nanomachines [1]. One of the three main approaches in building these so-called nanomachines is bio-hybrid nanomachines, nanomachines that are built based on simple biological entities (e.g. programmed cells, bacteria). MC focuses on systems that are either being used in biological systems or inspired by them instead of utilizing well-known communication systems such as electromagnetic wave based communication because these methods are envisioned to be compatible with bio-hybrid nanomachines. Also, these systems can be used as a transmission solution for communication between nanomachines and actual living cells, which are crucial for certain applications like interaction between prosthetic smart limbs and the nervous system.

Recently, numerous communication systems have been proposed in the literature in the context of MC. Among these systems, Ion signaling [18], microtubule networks [5], and Communication via Diffusion (CvD) [21] are developed based on actual biological inter- and/or intracell communication while others such as

E. Hart et al. (Eds.): BIONETICS 2011, LNICST 103, pp. 199–212, 2012.

bacteria based communication [4], pheromone signaling [8], Morphogenesis [14], and Fluorescence Resonance Energy Transfer (FRET) [13] are methods inspired from well-known biological or physical concepts. Being one of the first introduced methods, currently CvD has received much interest in the literature and has become one of the promising communication systems in MC.

The main idea behind the CvD system is the usage of certain molecules, called messenger molecules, as the information carriers between two nanomachines residing in close-to-medium proximity to each other in an aqueous environment (Figure 1). The system is composed of five key processes as encoding, transmission, propagation, reception, and decoding [1,21]. First, data is encoded upon one or several properties (e.g. concentration level) of a molecule wave. Then, based on the selected encoding technique and the bit sequence in question, the transmitter releases a number of molecules in a time slotted fashion. These messenger molecules dissipate in the medium following the prevalent probabilistic diffusion dynamics in the environment. This propagation behavior is modeled by the well-known Brownian motion. Based on several parameters such as temperature of the environment, Stoke's radius of the molecule in question, and distance between the transmitting couple, some of these released molecules arrive at the receiver within a given period of time and trigger a reception event. This event is accomplished with the assistance of protein based structures called receptors in the cell membrane of the receiving nanomachine. Finally, based on the properties of the received molecule wave, the information is decoded and understood by the receiver.

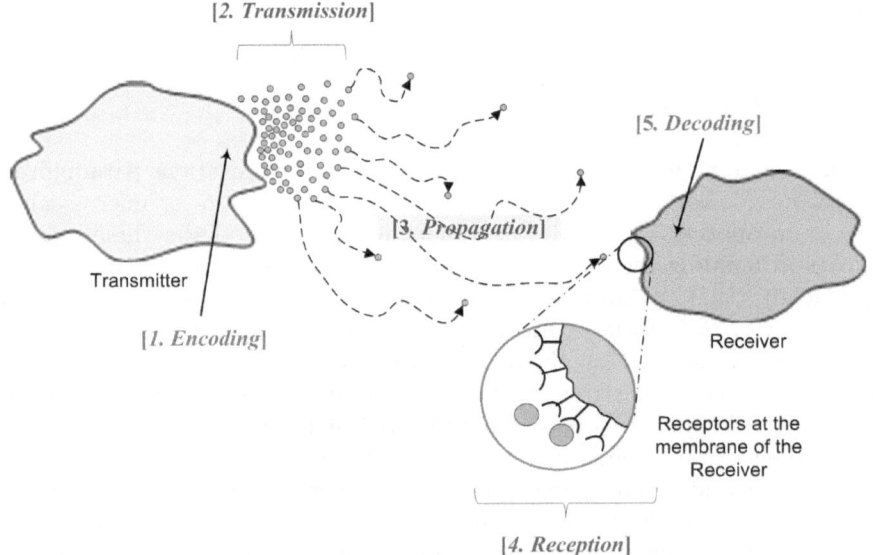

Fig. 1. Communication via Diffusion model

In the recent years, some elements of this aforementioned five process structure have found place in the literature. Most of these studies focus on the channel capacity and propagation dynamics of the CvD medium [3,19,2,17,15]. Some of these propagation process studies consider the probabilistic behavior of the channel as the transfer function of the system while others model it as a unique noise source inherent to a diffusion medium. According to the aforementioned studies on channel capacity and our own results in [10], it is shown that the reliability of the transmission diminishes exponentially with increasing transmission range while the average end-to-end delay increases exponentially. These results limit the effective communication range of the CvD systems to a few tens of micrometers; making it a solution for short-to-medium range inter-nanomachine communication.

A recent study shows the CvD channel to be a linear and time invariant [7] channel using a detailed simulator in a 2D environment. In [10], we explore the channel capacity of the CvD system constrained by the energy capabilities of the transmitting nanomachine. Our work has shown that the energy capabilities of an average human endocrine system cell, is theoretically enough for the transmitter to utilize a few hundred independent molecular channels simultaneously. Contrary to the popular models, Srinivas et al. study the CvD system in an environment with drift [20]. In this work, the authors consider the medium has a constant drift to a certain direction, similar to the flow of blood in the capillary system of animals.

Apart from the channel capacity issues, other studies show the effects of different modulation techniques on the overall performance of the system. Most of the studies in the literature use the received molecular concentration as the information carrying property of the wave, similar to the Amplitude Shift Keying (ASK) technique in classical communication literature [2,19,20]. Other modulation techniques have also been investigated in the literature. Garralda et al. describe the usage of Pulse Position Modulation (PPM) [7], Mahfuz et al. study the effects of Frequency Shift Keying (FSK), and we have developed a new modulation technique called Molecular Shift Keying (MSK) unique to the CvD medium while formalizing the ASK based techniques as Concentration Shift Keying (CSK) [12].

The transmission and reception processes have also been explored in the literature. In [17], Moore et al. propose a receiver structure that utilizes microtubules as exterior antennas that can capture propagating molecules that are far away from the receiver and thus increasing the transmission range. In a more recent work from the same authors, a directed antenna structure has been considered for the CvD system [16]. They propose to load the information inside a bacteria which follows certain molecules released from the target nanomachine. Based on the concentration of the molecules nearby, the bacteria, estimates the distance and direction of the receiver and steers itself accordingly. We have studied the effects of releasing molecules from different sides of a spherical transmitter in a 3D environment [11]. Our results show that, a release point selection technique based on the location of the receiver should be developed in order to avoid huge inefficiency in the CvD system.

Most of these studies on the CvD system focus on a single transmitter single receiver systems. However, when there are more communicating couples in the environment, additional issues arise and change the workings of the communication system. Thus, in order to develop a fully capable system for the CvD system in MC, we need to address these issues and design our communication system with these concerns in mind. An important one among these issues is the interference between closely placed transmitting couples in the same medium. When two or more transmitting pairs try to communicate simultaneously using the same modulation technique and same type of messenger molecules, their signals affect each other and reduce the signal to noise and interference ratio (SINR) of all nearby transmissions. In this paper, we study the effects of co-channel interference over the transmission performance with respect to several system parameters and evaluate the molecular reuse distance similar to the established frequency reuse range in the wireless electromagnetic communication literature.

The rest of the paper is organized as follows. In section 2, the interference types and the corresponding physical phenomena in CvD medium are explained. Section 3 describes the communication model and the channel capacity calculations developed to evaluate the co-channel interference in CvD. The simulation results that show the effects of several parameters on the co-channel interference are presented in Section 4. Lastly, Section 5 concludes the paper.

2   Interference Analysis in Communication via Diffusion

In communication systems, a given signal is affected by various sources while it propagates in the medium. All elements that affect a given signal are called interference to the signal. These effects can either be beneficial (constructive interference) or harmful (destructive interference) to the signal in question. The most important sources of interference are the intersymbol interference (ISI), adjacent channel interference (ACI), and co-channel interference (CCI). In any transmission, the signal is composed of a sequence of symbols, which upon aggregation form the whole data. ISI is defined by the interference effect of a symbol onto the successor symbols. When the transmitter sends a sequence of symbols in waveform, due to numerous reasons (e.g., multipath propagation, non-linear response from the channel, etc), the signal representing a given symbol can affect the subsequent symbols. ACI is the result of imperfect filtering in the transmitter, which results in the signal also having some components in the adjacent frequencies. Ideally, a transmitter should only transmit the signal at a given frequency, however due to wave forming limitations of filters, this property cannot be properly attained by a real-life transmitter circuit. Thus, a signal unavoidably causes interference to its adjacent channels. The last major interference source, the CCI, is the effect of concurrent transmissions onto each other in a given physical environment utilizing the same frequency. Since a signal propagates freely through the environment, when there are more than one transmitting couples in close proximity, some parts of a given transmitter's signal reach the other receivers. In addition to these interference sources, the signal is also affected by

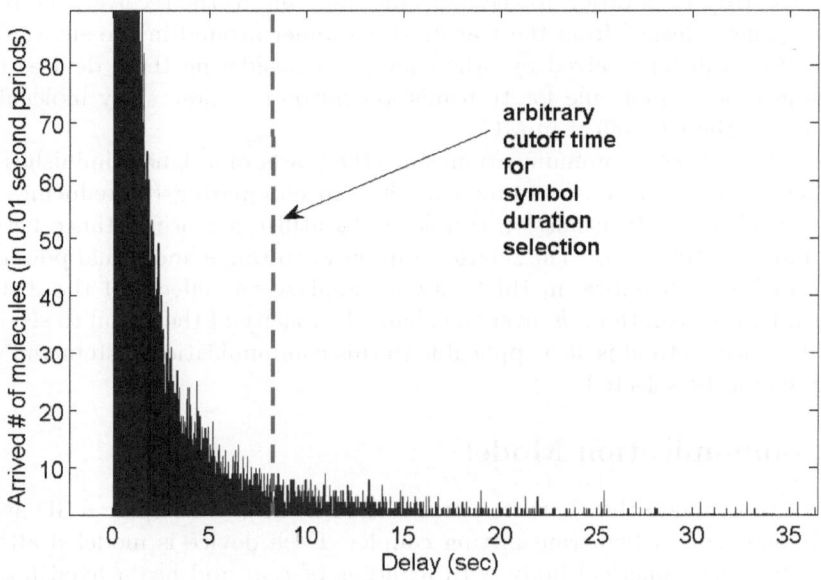

Fig. 2. Molecule arrival delay distribution at the receiver in CvD system

environmental background. However, these effects are not considered as a type of interference and are called noise. We have analyzed the effects of ISI in the CvD system in our previous work [10]. In this paper, while retaining the ISI effects, we focus on the CCI effects in the CvD system.

The prevalent force that affects the propagation of a signal in MC is the probabilistic behavior of the Brownian motion. While this behavior is fundamentally different from the well-known and well-studied deterministic medium of the electromagnetic wave based communication, the concept of interference still applies to MC. As we show in [12], the information can be encoded upon various properties of a molecular wave (e.g. concentration, molecule type, phase) to form a molecular signal. In this paper, we focus on the binary concentration shift keying (BCSK) method that is being widely used in the literature. Different from the BCSK system described in our previous work, in this paper we use a silence aware BCSK with two thresholds to enable the system to differentiate between silence in the channel and the bit values, instead of utilizing a single threshold to differentiate two bit values.

As seen in Figure 2, the concentration amount based received signal has a log-normal like distribution in MC. The amplitude of the signal is affected by the number of molecules released, and the variance is affected by the diffusion coefficient. Since the MC medium is inheritably slow in terms of propagation delay, the symbol duration should be selected to the left as much as possible in this distribution graph while including the spike part of the signal but leaving out some part of the long tail. However, the molecule arrival in this left out part of the tail affects the decoding process of the next symbol, and thus forms the ISI in the MC [10]. In addition to the molecules arriving to the receiver, the rest of the molecules released from the transmitter wander around in the environment and in time will be received by other devices. Considering these devices using the same type of molecule for transmission purposes, these stray molecules is the cause of the CCI effect in MC.

In electromagnetic communication, since the power of a signal diminishes with increasing range, the most common method in eliminating (or reducing) CCI among neighboring transmission couples is to utilize a reuse distance between the communicating pairs. The relationship between range and signal power also occurs in MC. Therefore, in this work we analyze the effects of the distance between two transmitters, h, over the channel capacity of the signal to see if the reuse distance method is also applicable to this communication system and show how it should be selected.

3   Communication Model

In order to evaluate the characteristics of CCI in MC, we employ a 3D system model composed of two transmitting couples. Each device is modeled after an identically sized spherical body with a radius of r_{cell} and has a fixed position in the topology. In each couple, the transmitter is separated from the receiver by a distance of d nanometers (Figure 3). Also, the distance between the two transmitters is denoted with h. The receivers are also apart from each other by the same amount.

A transmitter releases a given number of molecules depending on the bit value of the symbol (n_o for "0", and n_1 for "1"). Instead of a classical omnidirectional antenna used in EM communication and assuming a biological cell-like device, the whole nanomachine acts as the the antenna in MC. Hence, the release points of two cell-like nanomachines, each with a radius of r_{cell}, has a minimum distance of $2 * r_{cell}$ in between due to the volume of the devices. Thus, in addition to h, r_{cell} is also an important factor for the severity of CCI since it also affects the distance between the release points of the messenger molecules.

We assume each molecule has a spherical size with a radius of $r_{molecule}$ and propagates in the medium according to the Brownian motion whose diffusion coefficient is given as D. If a molecule collides with a receiver, we say "the molecule has hit the receiver", and the molecule is removed from the system since the ligand receptor at the receiver forms a chemical bond with the messenger molecule and the molecule is absorbed by the receiver [2]. It is assumed that all surface of the receiver is composed of receptors, which are able to bind with the messenger molecules. Unlike the receivers, the transmitters do not have receptors for this type of messenger molecules on their surface. Therefore, if a molecule hits a transmitter (either the one it is released from or the other), it bounces back from the transmitter. Since the messenger molecules operate in the low Reynolds number domain, they do not and can not have any inertia. Following this property, we model the bouncing molecule as canceling an illegal movement as if it did not happen at all [9].

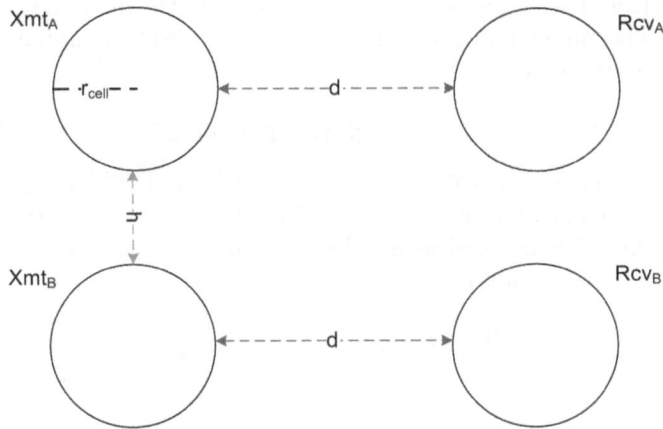

Fig. 3. Simulation model topology

After the molecules are released to the environment, some of them hit to a receiver fairly quickly while others hit after a long period of time and few wander around. Theoretically, if we wait indefinitely, every released molecule eventually hits a receiver. However, as stated in the previous section, in a communication system the information is encoded on a number of symbols and is expected to arrive at the receiver within a given duration, called symbol duration (t_s). According to our previous work, the selection of this symbol duration is heavily dependent on several parameters such as d and D. While a detailed analysis of symbol duration should be utilized, for the sake of simplicity we follow our previous method for choosing the symbol duration in this work. For each combination of d, r_{cell}, and h values used in the simulations, we take $100,000$ independent trials for the propagation of a single molecule. Among these trials we take the ones that hit the correct receiver and select the time required for $\alpha\%$ of them to hit the correct receiver as the symbol duration. According to our trials, we find out that the α value should be chosen close to 60, which enables reasonable values for both the symbol duration length and the number of unwanted surplus molecules to the subsequent symbol.

After finding out the appropriate t_s value, we re-run the molecule propagation trials for this duration and calculate the hitting probabilities of a single molecule to both of the receivers (P_{hit}^{R}, as hitting probability to the correct receiver and P_{hit}^{W} as the hitting probability to the wrong receiver). Then, we calculate the distribution of number of hitting molecules (N_c^{R}) when a given number of molecules (n) are released from the same point at the same time as

$$N_c^{R} \sim Binomial(n, P_{hit}^{R}(d, t_s)). \tag{1}$$

In addition to the molecules originating from the transmitter, other molecules may hit the receiver. Some of these molecules belong to the previous symbol of the signal while others originate from the current and previous symbols of the other transmitter. These sources act as ISI and CCI to the intended transmission, respectively. The number of molecules causing the ISI is denoted as N_p^R and follows a distribution as

$$N_p^R \sim Binomial(n, P_{hit}^R(d, 2t_s)) - Binomial(n, P_{hit}^R(d, t_s)). \tag{2}$$

The molecules causing the CCI is denoted as N_c^W and N_p^W for molecules belonging to the current and previous symbol of the other transmitter. Molecules coming from the other transmission follow similar distributions as the molecules from the main transmission as

$$N_c^W \sim Binomial(n, P_{hit}^W(d, t_s)) \tag{3}$$

and

$$N_p^W \sim Binomial(n, P_{hit}^W(d, 2t_s)) - Binomial(n, P_{hit}^W(d, t_s)). \tag{4}$$

Combining these four molecular arrival distributions in one symbol, the total number of molecules hitting the receiver in a t_s can be calculated as the summation of these sources

$$N = N_c^R + N_p^R + N_c^W + N_p^W. \tag{5}$$

Using a tri-state channel model for the silence aware BCSK, the symbol is decoded as silence if $N < \tau_0$, as "0" if $(N > \tau_0) \& (N < \tau_1)$, and as "1" if $(N > \tau_1)$.

Being a tri-state channel, the current symbol (s_c^m) and the other three symbols (previous symbol of the main transmission (s_p^m), previous symbol (s_p^o), and the current symbol (s_c^o) of the other transmission) can each take three values

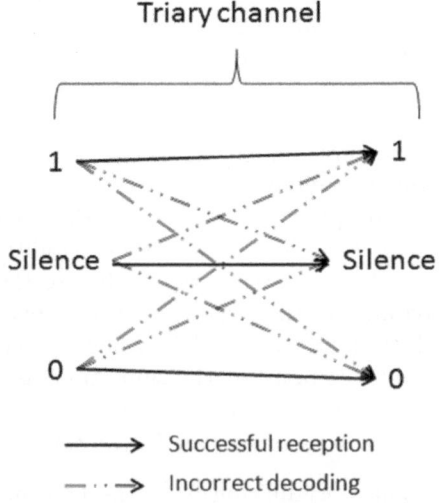

Fig. 4. Channel model for Silence Aware BCSK

(Figure 4). After substituting 0, n_0, and n_1 for n according to symbol values s_c^m, s_p^m, s_c^o, and s_p^o, the decoding probabilities and conditional channel capacities are calculated in all possible 81 cases. Summing up these conditional channel capacities with equally likely symbol values (each symbol has the same probability of being silence, 0, and 1), we calculate the overall channel capacity of the system using the well-known channel capacity formulation below

$$C = \max_{\tau} \sum_{y \in \{S,0,1\}} \sum_{x \in \{S,0,1\}} \mathbf{P}_{X,Y}(x,y) \log_2 \frac{\mathbf{P}_{X,Y}(x,y)}{\mathbf{P}_X(x)\mathbf{P}_Y(y)}. \tag{6}$$

4   Simulation Results

Based on the communication model above, we evaluate the effect of CCI in the CvD system with respect to different h values over two performance metrics, the probability of hitting to the receivers and the overall channel capacity. We run the simulations assuming a water-like environment in average body temperature with insulin hormone-sized messenger molecules. For precision in calculation, we choose the step time in the diffusion model as 0.001 seconds. The symbol durations are chosen based on $\alpha = 60$, and varying values of d. We use the average number of molecules emitted for each symbol $(n_0 + n_1)/2$ as the transmitter power, following the energy model in [10]. These simulation parameters are given in Table 1.

Table 1. Simulation parameters

Parameter	Value
Radius of messenger molecule ($r_{molecule}$)	2.5 nm [6]
Viscosity of the fluid (η)	0.001 $\frac{kg}{s \cdot m}$
Temperature (T)	$310°K$
Drag constant (b)	$5.391 \, 10^{-11} \frac{kg}{s}$
Diffusion coefficient (D)	79.4 $\frac{\mu m^2}{s}$
Step time (Δt)	0.001 s
Symbol duration (t_s) $d = 4\mu m$	0.213 s
Symbol duration (t_s) $d = 8\mu m$	0.949 s
Symbol duration (t_s) $d = 16\mu m$	4.064 s
Symbol duration (t_s) $d = 32\mu m$	17.391 s

First, we analyze the effect of h parameter over the hitting probabilities for different d values while setting r_{cell} to a moderate value ($5\,\mu m$). As seen in Figure 5, with the increase in h, P_{hit}^W decreases and eventually converges to zero while P_{hit}^R increases only slightly. Compliant to our previous works and other findings in the literature, P_{hit}^R decreases with increasing transmission range (d). However, P_{hit}^W does not show the same behavior. This is due to the fact that, when $d = 4\,\mu m$, the molecules have little space to move and most of them either

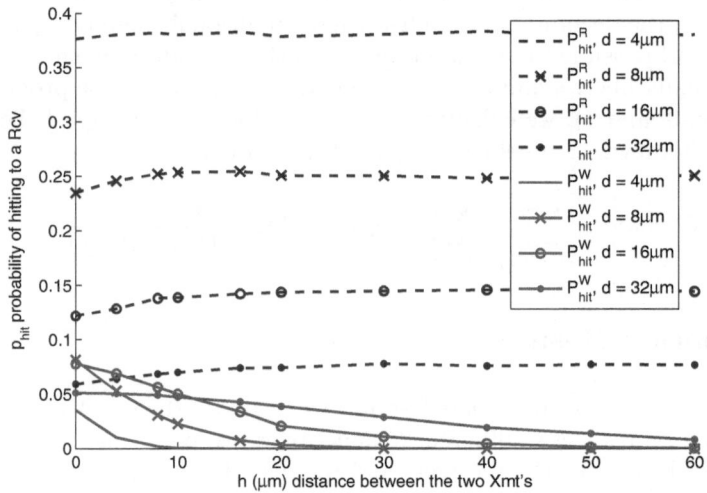

Fig. 5. Effect of h over the hitting probabilities with varying d

hit the correct receiver or dissipate in the environment. When d increases, the molecules move more freely in the environment and they have a higher chance of hitting the wrong receiver albeit the detrimental effect of increased range. As d further increases, both hitting probabilities decrease since the transmission range becomes the prevalent factor affecting the hitting probabilities.

Using these hitting probabilities, in Figure 6 we depict the channel capacities with varying h and d values while the average transmission power per symbol is 500 molecules (i.e. $n_0 = 300$ and $n_1 = 700$). When $d = 4\mu m$, the low P_{hit}^W values do not affect the overall channel capacity. Thus, increasing h has little benefit to the system. However, as d increases, CCI starts to affect the performance of the system, and increasing h becomes a good solution to mitigate CCI as in the electromagnetic wave-based wireless communication case.

Next, we set the transmission range as $16\,\mu m$ and analyze the hitting probabilities for different r_{cell} values. As seen in Figure 7, the cell radius heavily affects P_{hit}^R since hitting the receiver is proportional with the volume of the receiver in the 3D environment and the spherical volume decreases cubically with decreasing r_{cell} value. Similar to the previous graph, this behavior in P_{hit}^R is not reflected in the P_{hit}^W values. Although, the decrease in the volume of the target negatively affects P_{hit}^W, it is compensated by the decrease in the total distance from the molecular release point and the wrong receiver. As with the previous graph, increasing h decreases the CCI effects while increasing P_{hit}^R.

Note that the ratio between P_{hit}^R and P_{hit}^W for the $d = 32\,\mu m$ and $r = 5\,\mu m$ case in Figure 5 and $d = 16\,\mu m$ and $r = 2\,\mu m$ case in Figure 7 are very similar. Hence, there is a critical d value for a given r_{cell} after which P_{hit}^R and P_{hit}^W become very close to each other and the signal is heavily affected by the CCI due to the other transmission.

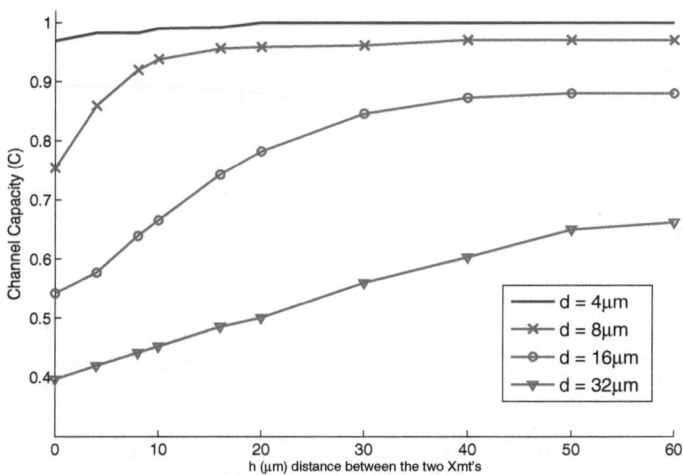

Fig. 6. Effect of h over Channel capacity with varying d

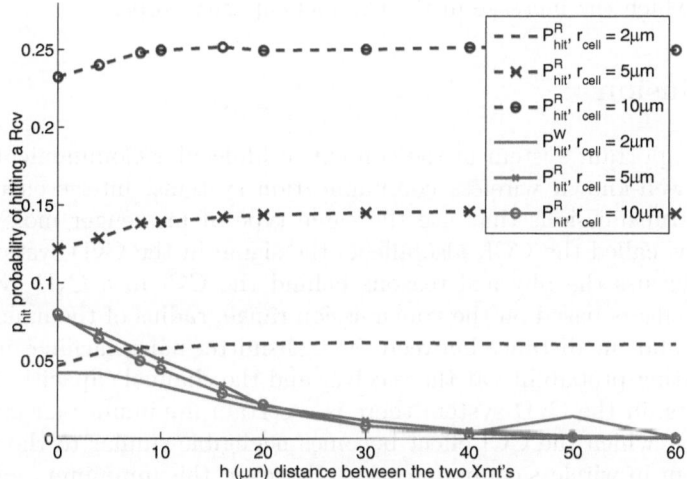

Fig. 7. Effect of h over the hitting probabilities with varying r_{cell}

Lastly, we evaluate the effect of average power over the channel capacity for different h values with a fixed transmission range of $16\ \mu m$s and r_{cell} of $5\ \mu m$s. As seen in Figure 8, at all power levels, increasing h increases the channel capacity up to a certain value depending on the transmission range. So, the positive effect of selecting a higher h value is not affected by the average symbol power, but the average symbol power determines the upper limit for the channel capacity

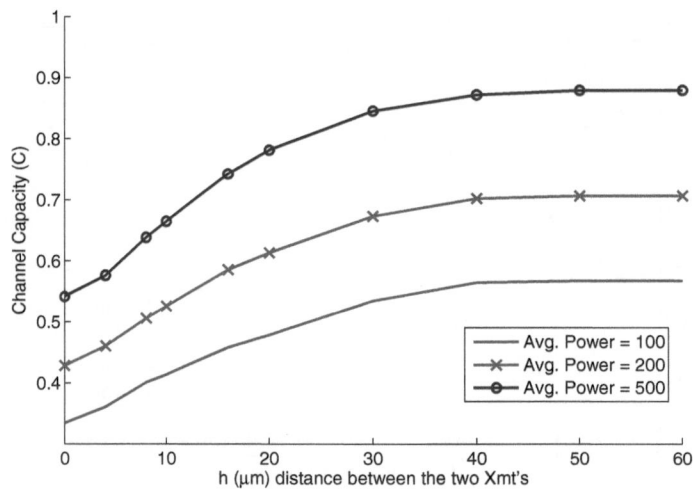

Fig. 8. Effect of h over channel capacity with varying average symbol power

when h is high enough. Note that, the average signal power does not change the point after which the increase in the channel capacity stops.

5   Conclusion

CvD is an important system in the context of Molecular Communication. Similar to the well-known wireless communication systems, interference between concurrent transmissions that use the same type of messenger molecules in a close vicinity, called the CCI, also affects the signal in the CvD system. In this paper, we discuss the physical reasons behind the CCI in a CvD system and evaluate its effects based on the transmission range, radius of the nanomachines in question, and the distance between the transmitter nanomachines in terms of molecule hitting probability at the receiver and the channel capacity. According to our results, in the CvD system there is a certain minimum molecular re-use distance after which the CCI effect becomes negligible similar to the frequency re-use concept in wireless communication. However, this minimum molecular re-use distance increases based on the distance between the molecule release points of the interfering transmitters, which is affected by both the distance between the transmitting nanomachines and their radii. As the future work, we plan to extend the CCI evaluation with increasing number of interfering transmission couples in the environment and finally find a molecular reuse distance calculation formula with given d, r_{cell}, D, and number of interfering transmission couples. Also we plan to extend our studies to the third important interference source in a communication system, the ACI, and combine them with our previous work regarding the ISI to come up with a thorough interference analysis for the CvD system.

Acknowledgment. This work has been supported by the State Planning Organization (DPT) of Republic of Turkey under the project TAM, with the project number 2007K120610.

References

1. Akyildiz, I.F., Brunetti, F., Blazquez, C.: Nanonetworks: A new communication paradigm. Elsevier Computer Networks 52(12), 2260–2279 (2008)
2. Atakan, B., Akan, O.B.: On Channel Capacity and Error Compensation in Molecular Communication. In: Priami, C., Dressler, F., Akan, O.B., Ngom, A. (eds.) Transactions on Computational Systems Biology X. LNCS (LNBI), vol. 5410, pp. 59–80. Springer, Heidelberg (2008)
3. Atakan, B., Akan, O.B.: Deterministic capacity of information flow in molecular nanonetworks. Elsevier Nano Communication Networks 1(1), 31–42 (2010)
4. Cobo, L.C., Akyildiz, I.F.: Bacteria-based communication in nanonetworks. Elsevier Nano Communication Networks 1(4), 244–256 (2010)
5. Enomoto, A., Moore, M., Nakano, T., Egashira, R., Suda, T.: A Molecular Communication System Using a Network of Cytoskeletal Filaments. In: 9th Nanotechnology Conference and Trade Show (NANOTECH 2006), pp. 725–728 (May 2006)
6. Freitas, R.A.: Nanomedicine, Vol. I: Basic Capabilities, 1st edn. Landes Bioscience (1999)
7. Garralda, N., Llatser, I., Cabellos-Aparicio, A., Pierobon, M.: Simulation-based evaluation of the diffusion-based physical channel in molecular nanonetworks. In: 2011 IEEE Conference on Computer Communications Workshops (INFOCOM WKSHPS), pp. 443–448. IEEE (April 2011)
8. Giné, L.P., Akyildiz, I.F.: Molecular Communication Options for Long Range Nanonetworks. Elsevier Computer Networks 53(16), 2753–2766 (2009)
9. Havlin, S., Ben-Avraham, D.: Diffusion in disordered media. Advances in Physics 36(6), 695–798 (1987)
10. Kuran, M.S., Birkan Yilmaz, H., Tugcu, T., Özerman, B.: Energy model for communication via diffusion in nanonetworks. Elsevier Nano Communication Networks 1(2), 86–95 (2010)
11. Kuran, M.S., Yilmaz, H.B., Tugcu, T.: Effects of routing for communication via diffusion system in the multi-node environment. In: 2011 IEEE Conference on Computer Communications Workshops (INFOCOM WKSHPS), pp. 461–466. IEEE (April 2011)
12. Kuran, M.S., Yilmaz, H.B., Tugcu, T., Akyildiz, I.F.: Modulation Techniques for Communication via Diffusion in Nanonetworks. In: 2011 IEEE International Conference on Communications (ICC), pp. 1–5. IEEE (June 2011)
13. Kuscu, M., Akan, O.B.: A nanoscale communication channel with fluorescence resonance energy transfer (FRET), pp. 425–430 (April 2011)
14. MacLennan, B.J.: Morphogenesis as a model for nano communication. Elsevier Nano Communication Networks 1(3), 199–208 (2010)
15. Mahfuz, M.U., Makrakis, D., Mouftah, H.T.: On the characterization of binary concentration-encoded molecular communication in nanonetworks. Elsevier Nano Communication Networks 1(4), 289–300 (2010)
16. Moore, M.J., Nakano, T.: Addressing by beacon coordinates using molecular communication. In: 2011 IEEE Conference on Computer Communications Workshops (INFOCOM WKSHPS), pp. 455–460. IEEE (April 2011)

17. Moore, M.J., Suda, T., Oiwa, K.: Molecular Communication: Modeling Noise Effects on Information Rate. IEEE Transactions on NanoBioscience 8(2), 169–180 (2009)
18. Nakano, T., Suda, T., Moore, M., Egashira, R., Enomoto, A., Arima, K.: Molecular Communication for Nanomachines Using Intercellular Calcium Signaling. In: 5th IEEE Conference on Nanotechnology (IEEE-NANO 2005), vol. 2, pp. 478–481 (July 2005)
19. Pierobon, M., Akyildiz, I.F.: A physical end-to-end model for molecular communication in nanonetworks. IEEE Journal on Selected Areas in Communications 28(4), 602–611 (2010)
20. Srinivas, K.V., Adve, R.S., Eckford, A.W.: Molecular communication in fluid media: The additive inverse Gaussian noise channel (December 2010)
21. Suda, T., Moore, M., Nakano, T., Egashira, R., Enomoto, A.: Exploratory Research on Molecular Communication between Nanomachines. In: Genetic and Evolutionary Computaion Conference (GECCO 2005). ACM (June 2005)

Channel Design and Optimization of Active Transport Molecular Communication

Nariman Farsad[1], Andrew W. Eckford[1], and Satoshi Hiyama[2]

[1] Dept. of Computer Science and Engineering, York University
4700 Keele Street, Toronto, ON, Canada M3J 1P3
`nariman@cse.yorku.ca, aeckford@yorku.ca`
[2] Research Laboratories, NTT DOCOMO Inc., Yokosuka, Kanagawa, Japan
`hiyama@nttdocomo.co.jp`

Abstract. In this paper, a guideline is provided for design and optimization of the shape of active transport molecular communication channels. In particular rectangular channels are considered and it is shown that for channels employing a single microtubule as the carrier of the information particles, the smaller the perimeter of the channel, the higher the channel capacity. Furthermore, it is shown that when channels with similar perimeters are considered, square-like channels achieve higher channel capacity for small values of time per channel use, while narrower channels achieve higher information rates for larger values of time per channel use.

Keywords: Molecular communication, microchannels, information theory.

1 Introduction

Molecular communication [1] is a new and emerging form of communication where molecules are used to carry information from a nano- or micro-scale transmitter to a nano- or micro-scale receiver. These information carrying molecules propagate in a fluidic environment we call fluidic microchannels, either passively or actively. In passive transportation, the information molecules diffuse in the channel via Brownian motion from the transmitter to the receiver. In active transport, the information molecules are transported as cargoes using a molecular motor system consisting of kinesins and microtubules.

As shown in [2,3,4], one advantage of active transport over passive transport is the higher degree of freedom in term of design parameters. For example, as shown in [2], parameters such as number of microtubules and the shape of the transmission zone can have a significant effect on the information rate of active transport propagation, where as the effect of such parameters on Brownian motion is nonexistent or minimal. Moreover, in [2] it was shown that active transport propagation can achieve higher information rates than Brownian motion. Therefore in this work we consider active transport propagation of information molecules, and propose elegant techniques for channel design optimization.

E. Hart et al. (Eds.): BIONETICS 2011, LNICST 103, pp. 213–223, 2012.

Notable works in this field include a general formulation of molecular communication as a timing channel under Brownian motion [5,6], an analysis of information transfer rates using molecular motors [7,8], mathematical channel models for continuous diffusion [9], binary concentration-encoded molecular communication [10], a simple model comparing the achievable information rates of passive transport using Brownian motion to that of active transport using microtubule filaments moving over a molecular motor track [2,11], a simple mathematical transport model for active transport propagation [3], and optimization of the transmission zone and vesicular encapsulation process [4]. In [12,13] basic channel design is considered for designing molecular sorters and rectifiers.

Our main contribution in this paper is design optimization of the rectangular channels employing active transport propagation. Typically laborious and time consuming methods relying on engineering intuitions and experimental trial-and-errors are used to design such channels. However in this work we use theoretical techniques to show that when the area of the channel is kept constant, the channel with smallest perimeter (i.e. square channel) maximizes the information transmission rate. When the perimeter is kept constant we show the surprising effect that the square channel achieves higher information rates at small values of time per channel use, while narrower channels achieve higher information rates for larger values of time per channel use.

2   Active Transport Molecular Communication Environment

Our molecular communication environment is similar to the ones in [2,11,14]. We use a rectangular propagation environment (with slightly rounded corners), consisting of a *transmission zone* and a *receiver zone*. Message-bearing particles originate at the transmission zone, and propagate until they arrive at the receiver zone. These particles are initially assumed to be anchored to the transmission zone until microtubule filaments, moving over molecular motor tracks that cover the whole environment, pick up and transport the information carrying particles from the transmission zone to the receiver zone.

As in [14], we assume that the microchannel environment is lined with static kinesin molecular motors, and that these motors cause microtubule filaments to propagate along their surface. The motion of the microtubule is largely regular (compared to pure Brownian motion), although the effects of Brownian motion cause random fluctuations. The loading and unloading mechanics are assumed to be the same as those proposed in [14]. The particles are anchored to the transmission zone through deoxyribonucleic acid (DNA) hybridization bonds, and do not move until they are picked up by a microtubule filament. The pickup and drop-off mechanisms are also assumed to be carried out through DNA hybridization bonds. This process is summarized in Figure 1 and the reader is referred to [14] for detailed explanation.

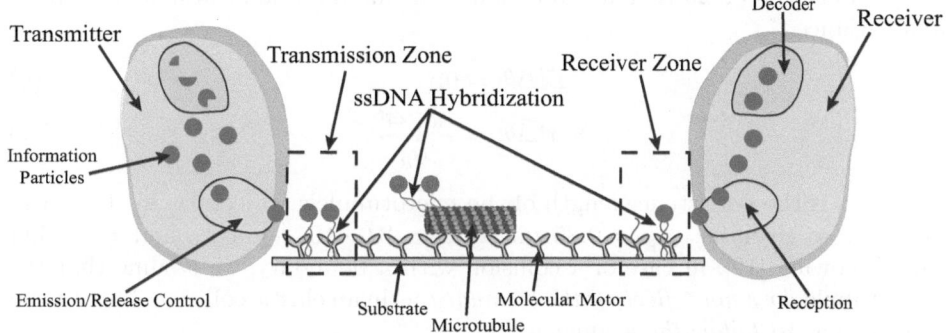

Fig. 1. Depiction of the communication environment

2.1   Simulating Active Transport

Although it has been demonstrated that it is possible to generate and study this
molecular communication system in wet labs [14,15], it is very laborious and difficult to study these systems from a communication perspective using laboratory
experiments. Therefore, previous works have relied on computer simulations to
study these systems. In [12,13] the motion of microtubule filaments over stationary kinesin molecular motors was simulated and in [2,11] a complete simulation
of the communication system was presented. In this work, for our simulations
we use the same techniques proposed in [2].

Since the microtubules move only in the x–y directions, and do not move in
the z direction (along the height of the channel), we consider a two-dimensional
simulation of microtubules for Δt time intervals. Given some initial position
(x_0, y_0) at time $t = 0$, for any integer $k > 0$, the motion of the microtubule is
given by the sequence of coordinates (x_i, y_i) for $i = 1, 2, \ldots, k$. Each coordinate
(x_i, y_i) represents the position of the microtubule's head at the end of the time
$t = i\Delta t$, where

$$x_i = x_{i-1} + \Delta r \cos\theta_i, \tag{1}$$
$$y_i = y_{i-1} + \Delta r \sin\theta_i. \tag{2}$$

In this case, the step size Δr at each step is an independent and identically
distributed (iid) Gaussian random variable with mean and variance

$$E[\Delta r] = v_{\mathrm{avg}}\Delta t, \tag{3}$$
$$\mathrm{Var}[\Delta r] = 2D\Delta t, \tag{4}$$

where v_{avg} is the average velocity of the microtubule, and D is the microtubule's
diffusion coefficient. The angle θ_i is no longer independent from step to step:
instead, for some step-to-step angular change $\Delta\theta$, we have that

$$\theta_i = \Delta\theta + \theta_{i-1}. \tag{5}$$

Now, for each step, $\Delta\theta$ is an iid Gaussian-distributed random variable with mean and variance

$$E[\Delta\theta] = 0, \tag{6}$$

$$\mathrm{Var}[\Delta\theta] = \frac{v_{\mathrm{avg}}\Delta t}{L_p}, \tag{7}$$

where L_p is the persistence length of the microtubule's trajectory. In [12], these values were given as $v_{\mathrm{avg}} = 0.85$ μm/s, $D = 2.0 \cdot 10^{-3}$ $\mu\mathrm{m}^2$/s, and $L_p = 111$ μm. Following [12], in case of a collision with a boundary, we assume that the microtubule *does not reflect off the boundary*, as in an elastic collision, but instead sets θ_i so as to *follow the boundary*.

In order to capture the loading process, in the simulations and the mathematical models, we use the grid loading structure proposed in [2]. For loading an information particle, the microtubule filament must drive close to the anchored particle. Therefore, we divide the transmission zone into a square grid, where the length of each square in the grid is the same as the diameter of the particles. We then distribute particles randomly and uniformly between the squares in the grid. If a microtubule enters a square which is occupied by a particle, we assume the microtubule loads that particle given it has an empty loading slot available. In general, based on laboratory experiments we assume that the microtubules can load multiple particles. For unloading, we assume all the loaded particles are unloaded as soon as a microtubule enters the receiver zone. Figure 2 shows a sample trajectory with the loading and unloading mechanism. Notice that the transmission and the receiver zones in this figure are rectangular strips across the width of the channel.

In the rest of this paper we assume the channel is rectangular with the transmission zone of the left and the receiver zone on the right side of the channel,

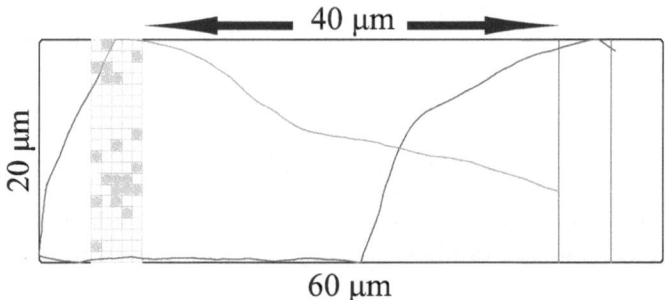

Fig. 2. A sample trajectory of active transport. The path of the microtubule is the line in the middle of the rectangular channel. The red portion corresponds to the path of the unloaded microtubule, while the green portion corresponds to the path of the loaded microtubule. The grid loading structure is on the left hand side and the unloading strip is on the right hand side. The cyan squares in the loading zone represent areas that contain an anchored particle which yellow squares represent areas with no particles.

as shown in Figure 3. We assume the width and length of the channel, as well as the size of the transmission and the receiver zones, are variable. However, for simplicity we assume the area of the transmission and the receiver zones are the same. Therefore, our rectangular channel environment can be characterized using three parameters: the width of the channel, the length of the channel, and the separation distance between the transmission zone and the receiver zone.

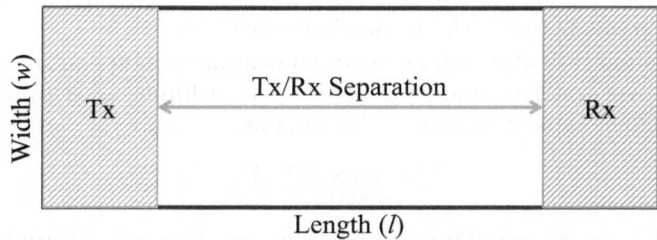

Fig. 3. The rectangular molecular communication environment with the transmission zone (blue striped area) on the left and the receiver zone (red striped area) on the right. The size of the transmission and receiver zones are assumed to be the same. The channel is characterized by its width, length and the separation between the transmission and the receiver zones.

3   Information Rate and Transport·Model

In this section we formulate the channel capacity of molecular communication channel as well as a transport model, all of which will be used to optimize the design of the channel in the next section. Previous work has considered molecular communication either as a *timing channel* problem (i.e., where information is encoded in the times when molecules are released) [5,8]; as an *inscribed matter* problem (i.e., where information is encoded by transmitting custom-made particles, such as specific strands of DNA) [16]; or as a *mass transfer* problem (i.e., a message is transmitted by moving a number of particles from the transmitter to the receiver) [2,8,10]. In this paper, we consider information transmission as a mass transfer problem.

In the simplest possible conception of this scheme, the particles themselves are not information-bearing, and a message is conveyed in the *number* of particles released by the transmitter. For example, if a maximum of three particles may be used, from a traditional communication system perspective, we may form messages two bits long (i.e., $\log_2 4$): "00" for 0 particles, "01" for 1 particle, "10" for 2 particles, and "11" for 3 particles. However, this message might not be perfectly conveyed to the receiver: given a time limit T for the communication session, it is possible that some of the particles will not arrive at the receiver after T has elapsed.

Since we have assumed that messages are encoded in the number of particles, let $X \in \mathcal{X}$ represent the number of particles released into the medium by the transmitter (i.e. the symbol transmitted by the source), $Y \in \mathcal{Y}$ represent the number that arrive at the destination (i.e. the symbol received at the destination) once T seconds have elapsed, and x_{max} be the maximum number of particles the transmitter can release per channel use. In other words, the set of possible transmission symbols and the set of possible received symbols are $\mathcal{X} = \mathcal{Y} = \{0, 1, 2, \cdots, x_{max}\}$. From the channel's perspective, $X \in \mathcal{X}$ is a discrete random variable given by probability mass function (PMF) $f_X(x)$, $Y \in \mathcal{Y}$ is also a discrete random variable given by PMF $f_Y(y)$, and a channel use is defined as T second intervals between the transmission releases.

The maximum rate at which any communication system can *reliably transmit information* over a noisy channel is bounded by a limit called *channel capacity* [17]. The channel capacity can be calculated as,

$$C = \max_{f_X(x)} I(X;Y), \tag{8}$$

where $I(X;Y)$ is the mutual information between X and Y. Mutual information is defined as

$$I(X;Y) = E\left[\log_2 \frac{f_{Y|X}(y|x)}{\sum_x f_{Y|X}(y|x) f_X(x)}\right], \tag{9}$$

where, $f_{Y|X}(y|x)$ represents the probability of receiving symbol y at the destination, given that symbol x was transmitted by the source; $f_X(x)$ represents the probability of transmitting symbol x; and $E[\cdot]$ represents expectation.

In general, if PMF $f_{Y|X}(y|x)$ is known, we can calculate mutual information for any $f_X(x)$. However, in order to calculate the channel capacity, we need to find the PMF $f_X(x)$ that maximizes mutual information. We can use the Blahut-Arimoto algorithm [18,19] to find the PMF $f_X(x)$ such that, given $f_{Y|X}(y|x)$, mutual information is maximized. Therefore, if PMF $f_{Y|X}(y|x)$ is known, we can calculate the channel capacity of the molecular communication system in a straightforward manner.

In [3,4], a simple transport model is presented to estimate the PMF $f_{Y|X}(y|x)$ in a computationally efficient manner. This transport model also provides the necessary insight into the underlying mechanics effecting $f_{Y|X}(y|x)$. According to this model, the PMF $f_{Y|X}(y|x)$ depends on two random variables: a Bernoulli random variable V_i, representing the event that the ith square is visited in a single microtubule trip from the transmitter to the receiver, and a random variable K representing the number of microtubule trips between the transmitter and the receiver during the time interval T. In [3,4] the random variable V_i was considered for optimization of the shape of the transmission zone. In this work, we study the effects of channel size on these random variables and use them to optimize the design.

4   Design Analysis

In this section, we consider the design analysis of rectangular microchannels to maximize their channel capacity. In particular, we consider properties of rectangular channels such as area and perimeter as optimization criterion. We separate the effects of area and perimeter by considering two scenarios: 1) we keep the area constant and change the perimeter; and 2) we keep the perimeter constant and change the area of the rectangular channel. Table 1 summarizes the sample channels with constant area, and Table 2 shows the sample channels with constant perimeter.

Table 1. Channels with constant area and different perimeters

Dimensions ($w \times l$)	Tx/Rx Separation	Area (A)	Perimeter (P)
5 μm $\times$ 240 μm	160 μm	1200 μm^2	490 μm
10 μm $\times$ 120 μm	80 μm	1200 μm^2	260 μm
20 μm $\times$ 60 μm	40 μm	1200 μm^2	160 μm
40 μm $\times$ 30 μm	20 μm	1200 μm^2	140 μm

Table 2. Channels with constant perimeter and different areas

Dimensions ($w \times l$)	Tx/Rx Separation	Area (A)	Perimeter (P)
10 μm $\times$ 70 μm	30 μm	700 μm^2	160 μm
20 μm $\times$ 60 μm	40 μm	1200 μm^2	160 μm
30 μm $\times$ 50 μm	36 μm	1500 μm^2	160 μm
40 μm $\times$ 40 μm	30 μm	1600 μm^2	160 μm

In the next two subsections we use these two sets of channels to analyse the effects of channel size on the number of microtubule trips and the probability that squares in the grid loading area are visited.

4.1   Constant Area

If the area of the rectangular channel is kept constant, increasing the perimeter yields longer and narrower channels. Since the microtubules mostly follow the walls of the channel, the larger the perimeter the smaller the average value of K (i.e. the average number of microtubule trips during time interval T). This effect can be seen by comparing the probability density function (PDF) of the 5 μm $\times$ 240 μm and the 40 μm $\times$ 30 μm channels as shown in Figure 4 for time per channel use of 750 seconds. There are zero microtubule trips for the longer and narrower channel (5 μm $\times$ 240 μm), while there are between 1 and 4 microtubule trips for the square-like channel (40 μm $\times$ 30 μm).

In contrast, as the channel gets narrower the probability that a square in the grid is visited is increased, because a larger number of squares are close to the walls. In [3,4] it was shown that the squares that are closer to the walls of the channel have a much higher probability of being visited in a single microtubule trip. However, the number of trips a microtubule takes between the transmitter and the receiver over time duration T is a more dominant factor. This is because

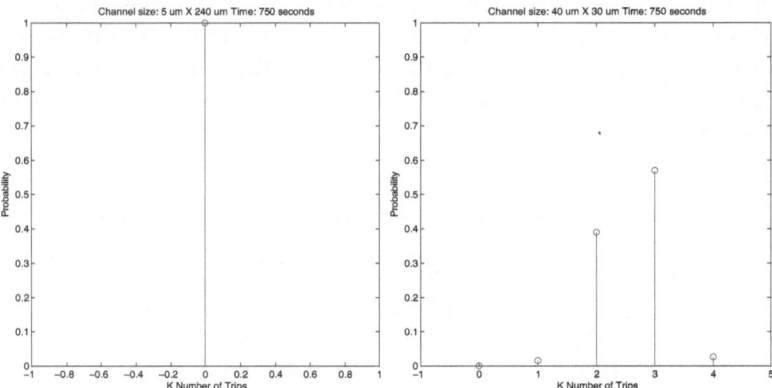

Fig. 4. Probability distribution function of K

the molecular communication is a mass transfer problem, which depends on the microtubules. Therefore, we would expect that keeping the area constant the channel is optimized when the perimeter is smallest (i.e. the channel is a square). This effect will be shown in the results section.

4.2   Constant Perimeter

If the perimeter of the rectangular channel is kept constant, the average number of trips a microtubule travels during time interval T seconds are almost identical regardless of the dimensions of the channel. This effect is due to the fact that microtubules follow the walls of the channel with a high probability, as shown in [4]. However, as the channel becomes narrower the number of trips becomes more deterministic, while as the channel gets closer to a square the number of trips become more random (i.e. the variance of K decreases as the channel gets narrower, while the variance increases when the channel is closer to a square). This effect can be seen in Figure 5 where the PDF of K is plotted over time duration of 750 seconds for the 10 μm $\times$ 70 μm and the 40 μm $\times$ 40 μm channels. While that average number of trips in both channels is close to 2, for the variance of K is lower for the narrower channel compared to the square channel.

Another interesting and surprising effect is that for small values of time duration per channel use, the higher variance of the square channels are constructive. For example, from Figure 6 we can see that for time durations of 250 seconds, there is a higher chance of a single microtubule trip between the transmitter and the receiver for a square-like channel compared to a narrower channel. Therefore, for small time durations we would expect the square-like channels to be better and for larger time durations the narrower channel.

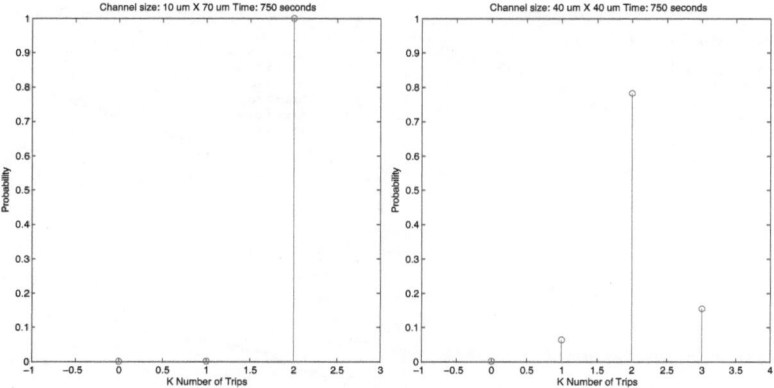

Fig. 5. Probability distribution function of K

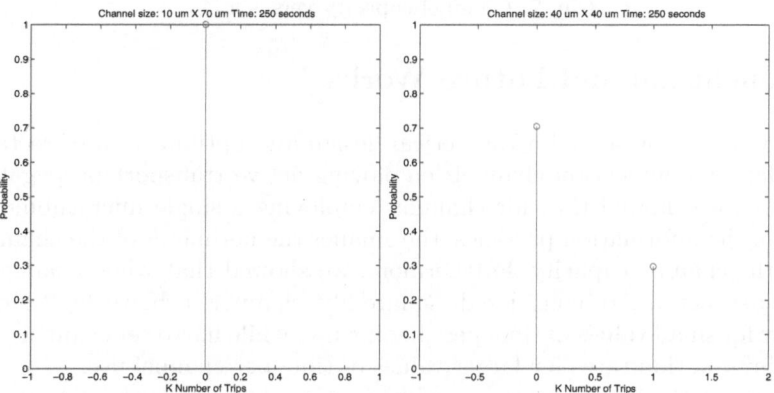

Fig. 6. Probability distribution function of K

5   Results

Figure 7 shows the channel capacity in bits versus x_{max} for different values of time per channel use. From the results we can see that channels with smaller perimeter generally achieve higher information rates because the microtubules can travel a larger number of times between the transmitter and the receiver. Moreover, we can see that for small time per channel use values when the channels have similar perimeters, the square-like channels perform better because there is a higher chance that the microtubules travels once between the transmitter and the receiver. However when perimeter is kept constant, as time per channel use increases, the narrower channels achieve higher information rates.

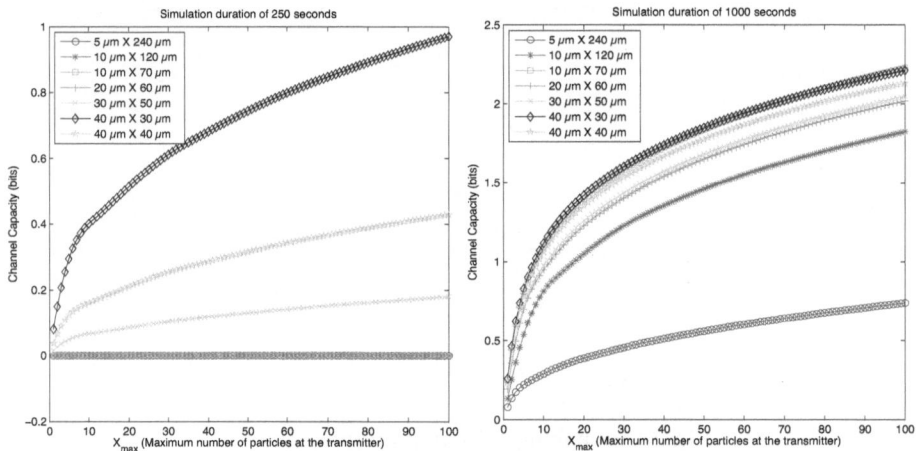

Fig. 7. Channel capacity versus x_{max}

6   Conclusion and Future Works

In this work we presented a theoretical design and optimization of rectangular molecular communication channels employing active transport propagation. In particular, we showed that for channels employing a single microtubule as the carrier of the information particles, the smaller the perimeter of the channel, the higher the channel capacity. Furthermore, we showed that when channels with similar perimeters are considered, square-like channels achieve higher channel capacity for small values of time per channel use, while narrower channels achieve higher information rates for larger values of time per channel use.

As part of the future work we will consider channels with multiple microtubules. In these channels as the area of the channel increases, the number of microtubules assisting in transmission increases. Using the results from this paper, and the fact that lager channel areas can accommodate larger number of microtubules, we want to maximize area while we minimize the perimeter. Solving this optimization problem will result in an optimal channel. Because circle has the largest area to perimeter ratio, we will also consider non-rectangular channels such as circular and ring-shaped channels.

References

1. Hiyama, S., Moritani, Y.: Molecular communication: harnessing biochemical materials to engineer biomimetic communication systems. Nano Communication Networks 1(1), 20–30 (2010)
2. Farsad, N., Eckford, A.W., Hiyama, S., Moritani, Y.: Information rates of active propagation in microchannel molecular communication. In: Proc. of the 5th International ICST Conference on Bio-Inspired Models of Network, Information, and Computing Systems, Boston, MA, p. 7 (2010)

3. Farsad, N., Eckford, A.W., Hiyama, S., Moritani, Y.: A simple mathematical model for information rate of active transport molecular communication. In: Proc. IEEE INFOCOM Workshops, Shanghai, P. R. China, pp. 473–478 (2011)
4. Farsad, N., Eckford, A.W., Hiyama, S., Moritani, Y.: Quick system design of vesicle-based active transport molecular communication by using a simple transport model. Nano Communication Networks (2011), doi:10.1016/j.nancom.2011.07.003
5. Eckford, A.W.: Nanoscale communication with brownian motion. In: Proc. of the 41st Annual Conference on Information Sciences and Systems, Baltimore, MD, pp. 160–165 (2007)
6. Atakan, B., Akan, O.: An information theoretical approach for molecular communication. In: Proc. of the 2nd International Conference on Bio-Inspired Models of Network, Information and Computing Systems, Budapest, Hungary, pp. 33–40 (2007)
7. Moore, M.J., Suda, T., Oiwa, K.: Molecular communication: modeling noise effects on information rate. IEEE Transactions on NanoBioscience 8(2), 169–180 (2009)
8. Eckford, A.W.: Timing Information Rates for Active Transport Molecular Communication. In: Schmid, A., Goel, S., Wang, W., Beiu, V., Carrara, S. (eds.) Nano-Net 2009. LNICST, vol. 20, pp. 24–28. Springer, Heidelberg (2009)
9. Pierobon, M., Akyildiz, I.F.: A physical end-to-end model for molecular communication in nanonetworks. IEEE Journal on Selected Areas in Communications 28(4), 602–611 (2010)
10. Mahfuz, M.U., Makrakis, D., Mouftah, H.T.: On the characterization of binary concentration-encoded molecular communication in nanonetworks. Nano Communication Networks 1(4), 289–300 (2010)
11. Eckford, A.W., Farsad, N., Hiyama, S., Moritani, Y.: Microchannel molecular communication with nanoscale carriers: Brownian motion versus active transport. In: Proc. of the IEEE International Conference on Nanotechnology, Seoul, South Korea, pp. 854–858 (2010)
12. Nitta, T., Tanahashi, A., Hirano, M., Hess, H.: Simulating molecular shuttle movements: Towards computer-aided design of nanoscale transport systems. Lab on a Chip 6(7), 881–885 (2006)
13. Nitta, T., Tanahashi, A., Hirano, M.: In silico design and testing of guiding tracks for molecular shuttles powered by kinesin motors. Lab on a Chip 10(11), 1447–1453 (2010)
14. Hiyama, S., Gojo, R., Shima, T., Takeuchi, S., Sutoh, K.: Biomolecular-motor-based nano- or microscale particle translocations on DNA microarrays. Nano Letters 9(6), 2407–2413 (2009)
15. Hiyama, S., Moritani, Y., Gojo, R., Takeuchi, S., Sutoh, K.: Biomolecular-motor-based autonomous delivery of lipid vesicles as nano- or microscale reactors on a chip. Lab on a Chip 10(20), 2741–2748 (2010)
16. Cobo, L.C., Akyildiz, I.F.: Bacteria-based communication in nanonetworks. Nano Communication Networks 1(4), 244–256 (2010)
17. Cover, T.M., Thomas, J.A.: Elements of Information Theory, 2nd edn. Wiley-Interscience (2006)
18. Blahut, R.: Computation of channel capacity and rate-distortion functions. IEEE Transactions on Information Theory 18(4), 460–473 (1972)
19. Arimoto, S.: An algorithm for computing the capacity of arbitrary discrete memoryless channels. IEEE Transactions on Information Theory 18(1), 14–20 (1972)

Applying Bees Algorithm for Trust Management in Cloud Computing

Mohamed Firdhous[1,2], Osman Ghazali[1], and Suhaidi Hassan[1]

[1] InterNetWorks Research Group, School of Computing, College of Arts & Sciences,
Universiti Utara Malaysia, Malaysia
mfirdhous@internetworks.my, {osman,suhaidi}@uum.edu.my
[2] Faculty of Information Technology, University of Moratuwa, Sri Lanka

Abstract. Cloud computing is considered the new paradigm in computing that would make computing a utility. Once the cloud computing becomes available widespread, many service providers would market their services at different qualities and prices. When this happens, the customers would be required to select the right service provider who could meet their anticipated quality. A trust management system would identify the quality of service providers and help customers to choose the right provider. Designing a trust management system is a difficult task, as it requires the consideration of several attributes both local and external to the system. In this paper, the authors propose that the Bees Algorithm that was used to solve issues in diverse fields could be successfully adapted to address the trust issue in the cloud computing system. The authors justify their proposition based on the comparative study carried out on cloud computing and the bees environments.

Keywords: cloud computing, trust management, bees algorithm.

1 Introduction

Cloud computing has been considered the new paradigm that envisages to change how the computing resources including hardware, software and services have been purchased and used [1]. Until the development of cloud computers, computing resources have been either purchased outright or leased from data centres. Purchasing computing resources outright or leasing the resources from data centre providers require payments for these resources irrespective of their usage. Cloud computing changes all these and make computing a utility similar to electricity, water, gas and telephony where users would pay only for the resources consumed [1]. Currently there are several commercial cloud service providers selling their services to a wider audience and there will be many more in time to come. These providers market different services at different levels of quality and price. Hence it is necessary for customers to identify the right provider before making a commitment. Trust systems can play a major role in identifying the service providers and their quality. In this paper the authors propose that the Bees Algorithm that has been used to solve optimization problems commonly

E. Hart et al. (Eds.): BIONETICS 2011, LNICST 103, pp. 224–229, 2012.

known as NP-hard problems can be adapted to solve the trust issue in cloud computing.

This paper consists of five sections where Section 1 introduces the paper and the topic while Section 2, and 3 concentrate n cloud computing and related Work respectively. Section 4 analyses the similarity between the environments where bees carry out their food foraging and the cloud computing environment. Finally Section 5 concludes the paper stating how the Bees Algorithm can be adapted to solve the trust management problem in cloud computing.

2   Cloud Computing

Cloud computing has created a new paradigm shift in computing by making the cloud resources available on the Internet as services. Consumers may be able to access computing resources like hardware, software, communication and services as utilities and pay for only what is used similar to other utilities. Hence cloud computing has now been commonly known as the 5th utility [1]. Cloud providers implement the services on virtualized platforms where these virtual systems can be brought up and destroyed based on user demand within short times [2]. Hosting systems on virtual servers helps the providers with the flexibility for selling the same resources to multiple clients reducing the cost of these resources per client while increasing the utilization of them. Purchasing cloud services in place of dedicated resources helps the businesses to move the capital expenditure on these resources to operational expenditure that can be spread over a longer period.

Cloud system can be divided into five layers in two groups as shown in Fig. 1 [3]. The physical hardware layer is the workhorce that powers the cloud system in terms of CPU, memory, storage, communication interface etc. The physical hardware layer can be implemented using clusters, data centres, work group or desktop computers [4]. The virtual hardware layer hosted on top of the physical hardware is the one that makes the cloud system different from other traditional hosting services. The virtualized systems can be brought up and down at the demand of the clients on the fly without affecting other virtual systems running on the same physical hardware. The middleware that virtualizes the physical hardware provides the necessary operating environment for applications maintaining isolation, quality of service and security [4]. Cloud service layers are divided into Infrastructure as Service (IaaS), Platform as a Service (PaaS) and Software as a Service (SaaS) layers [4]. Provision of processing power, storage, database etc., on the virtual platform is known as IaaS. The virtual systems made available to clients as IaaS can be treated similar to real systems and users can install the operating system and applications of their choice on them. The PaaS installed on top of the IaaS provides the platform for complete software development life cycle. PaaS offerings usually comprises programming languages, testing tools and Application programming Interface (API) to access these tools. Thus complete

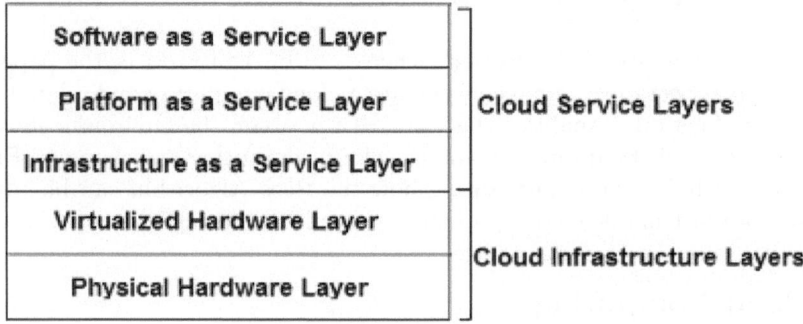

Fig. 1. Layered Architecture of Cloud Computing System [3]

applications can be developed, tested and deployed using PaaS. SaaS provides a radically new platform for marketing software tools and applications as services. This helps developers to focus totally on the application development without getting distracted by system installation and maintenance tasks. Applications developed and hosted on the Internet can now be purchased by clients and pay only for the services accessed based on the access time. This frees the customers from purchasing, installing and managing applications on site[5].

3   Related Work

3.1   Trust and Trust Management

Trust is considered to be an important factor when entities interact with each other involving uncertainty and dependency [6]. Trust has its origin in social sciences and studied by researchers in diverse fields such as psychology, sociology, economics and technology. As different researchers have approached trust from different angles depending on the field of specialization they are in, there was no common agreement or definition for what trust is? Though there are multiple definitions and approaches for dealing with trust, computer scientists can benefit from all these studies carried out on different fields [7]. Trust management systems have been implemented in distributed computing systems including peer to peer systems, grid computing, cluster computing, sensor networks and cloud computing [3].

3.2   Trust Management in Cloud Computing

Several researchers have attempted to build trust management systems for cloud computing [8,9,10,11,12]. Though most of these proposals had tried to address the trust in cloud computing, they lack a sound theoretical foundation. Firdhous et al. have proposed two trust models using Quality of Service (QoS) as the basis for building trust [3,13]. Though these two proposals lay a foundation for computing and adapting the trust value by continuously monitoring the performance of the service provider, they are based only on one QoS parameter namely the response time of the server.

3.3   Bees Algorithm

The Bees Algorithm is a population based search and optimization algorithm developed based on the food foraging behaviour of honey bees. Honey bees are social insects similar to ants, termites and wasps that live in large colonies and carry out their jobs with precise coordination [14]. In a typical beehive there are 20,000 - 100,000 bees which are organized in a hierarchical structure with the queen bee on top followed by female worker bees and male bees [15]. The worker bees are responsible for collecting nectar to produce honey and bee wax. Honey bees travel long distances looking for flower patches suitable for collecting nectar [16].

At the beginning of the foraging process, scout bees go out searching for suitable flower patches. These scout bees travel randomly moving from one patch to another. On their return to the hive, the ones those found a flower patch above certain quality deposit the nectar and perform the waggle dance. This waggle dance conveys the required information to other bees for exploring the flower patches [17]. Depending on the quality of the flower patches, number of bees are assigned to each flower patch for foraging. Large number of bees are assigned to flower patches with large amount of nectar while smaller flower patches are foraged by fewer bees. The scout bees that discovered the flower patch will lead other bees to the patch for further foraging. The food level of the patch is continuously monitored during the harvesting process as it is important for future harvesting [18].

The Bees Algorithm starts by searching the neighbourhood along with random search that is suitable for both combinatorial and functional optimization [17]. Several variations of the Bees Algorithm have been developed and published for solving various complex optimization problems [18,19,20,21,22].

4   Adoption of Bees Algorithm for Cloud Computing

If a detailed comparison between the cloud environment and the natural environment where bees forage for food, it is possible to see a remarkable similarities between these environments. Table 1 presents a summary of such comparison.

From Table 1, it can be seen that there is a remarkable similarity between cloud computing and bees environment. So Bees Algorithm could be applied successfully to the cloud computing system to address the trust management issue. By applying the Bees Algorithm for the cloud system, the trust of a service provider can be computed combining multiple parameters of the system and the environment. Also, it could be possible to optimize the performance of the system using the same Bees Algorithm, by allocating suitable number of clients to each server system.

Table 1. Comparison of Cloud Computing Environment with Bees Environment

Bees Environment	Cloud Environment
Flowers patches	Server systems
Different flower patches	Different types of services
Number of foraging bees	Number of clients
Duration of foraging	Duration of service
Scout bees	Software agents
Information from waggle dance	Information from agents
Direction	Route
Distance	Access time
Quality of nectar	Quality of service
Size of the patch	Capacity of the server system
Monitoring food level	Continuous update

5   Conclusions

In this paper, the authors present the idea that the Bees Algorithm could be adapted to solve the trust management issue in the cloud computing. The authors also present a detailed comparison between the cloud computing environment and the bees environment. The findings of the comparison support the conclusion of the authors that the Bees Algorithm could be successfully adapted to use in the cloud computing systems.

References

1. Buyya, R., Chee, S.Y., Venugopal, S., Broberg, J., Brandic, I.: Cloud Computing and Emerging IT Platforms: Vision, Hype, and Reality for Delivering Computing as the 5th Utility. Future Generation Computer Systems 26(6), 599–616 (2009)
2. Zaman, S., Grosu, D.: Combinatorial Auction-based Allocation of Virtual Machine Instances in Clouds. In: 2nd IEEE International Conference on Cloud Computing Technology and Science (CloudCom), pp. 127–134. IEEE Press, New York (2010)
3. Firdhous, M., Ghazali, O., Hassan, S.: A Trust Computing Mechanism for Cloud Computing with Multilevel Thresholding. In: 6th International Conference on Industrial and Information Systems (ICIIS), pp. 457–461. IEEE Press, New York (2011)
4. Buyya, R., Pandey, S., Vecchiola, C.: Cloudbus Toolkit for Market-Oriented Cloud Computing. In: Jaatun, M.G., Zhao, G., Rong, C. (eds.) CloudCom 2009. LNCS, vol. 5931, pp. 24–44. Springer, Heidelberg (2009)
5. Prodan, R., Ostermann, S.: A Survey and Taxonomy of Infrastructure as a Service and Web Hosting Cloud Providers. In: 10th IEEE/ACM International Conference on Grid Computing, pp. 17–25. IEEE Press, New York (2009)
6. Aljazzaf, Z.M., Perry, M., Capretz, M.A.M.: Online Trust: Definition and Principles. In: 5th International Multi-Conference on Computing in the Global Information Technology (ICCGI), pp. 163–168. IEEE Press, New York (2010)

7. Firdhous, M., Ghazali, O., Hassan, S.: Trust Management in Cloud Computing A Critical Review. International Journal on Advances in ICT for Emerging Regions (ICTer) 4(2), 24–36 (2011)
8. Khan, K.M., Malluhi, Q.: Establishing Trust in Cloud Computing. IT Professional 12(5), 20–27 (2010)
9. Song, Z., Molina, J., Strong, C.: Trusted Anonymous Execution: A Model to RaiseTrust in Cloud. In: 9th International Conference on Grid and Cooperative Computing (GCC), pp. 133–138. IEEE Press, New York (2010)
10. Sato, H., Kanai, A., Tanimoto, S.: A Cloud Trust Model in a Security Aware Cloud. In: 10th IEEE/IPSJ International Symposium on Applications and the Internet (SAINT), pp. 121–124. IEEE Press, New York (2010)
11. Wang, T.F., Ye, B.S., Li, Y.W., Yang, Y.: Family Gene based Cloud Trust Model. In: International Conference on Educational and Network Technology (ICENT), pp. 540–544. IEEE Press, New York (2010)
12. Shen, Z., Li, L., Yan, F., Wu, X.: Cloud Computing System Based on TrustedComputing Platform. In: International Conference on Intelligent Computation Technology and Automation (ICICTA), pp. 942–945. IEEE Press, New York (2010)
13. Firdhous, M., Ghazali, O., Hassan, S.: A Trust Computing Mechanism for Cloud Computing. In: 4th ITU-T Kaleidoscope Academic Conference: The Fully Networked Human? Innovations for Future Networks and Services. IEEE Press, New York (2011)
14. Wilson, E.O.: Sociobiology: The New Synthesis. Harvard University Press, Boston (2000)
15. Lazaryan, D.S., Sotnikova, E.M., Evtushenko, N.S.: Standardization of Bee Brood Homogenate Composition. Pharmaceutical Chemistry Journal 37(11), 614–616 (2003)
16. Needham, A.W.: The E-Book on Honey Bees (2010),
http://www.beginningbeekeeping.com/TheE-BookOnHoneyBees.pdf
17. Pham, D.T., Ghanbarzadeh, A., Ko, E., Otri, S., Rahim, S., Zaidi, M.: The Bees Algorithm A Novel Tool for Complex Optimisation Problems. In: Pham, D.T., Eldukhri, E.E., Soroka, A.J. (eds.) Proccedings of the 2nd I*PROMS Virtual International Conference, pp. 454–459. Elsevier, London (2006)
18. Mahmuddin, M.: Optimisation using Bees Algorithm on Unlabelled Data Problems. PhD Thesis, Cardiff University, Wales, UK (2008)
19. Pham, D.T., Afify, A., Koc, E.: Manufacturing Cell Formation using the Bees Algorithm. In: I*PROMS 2007 Innovative Production Machines and Systems Virtual Conference (2007)
20. Pham, D.T., Soroka, A.J., Ghanbarzadeh, A., Koc, E., Otri, S., Packianather, M.: Optimising Neural Networks for Identification of Wood Defects using the Bees Algorithm. In: IEEE International Conference on Industrial Informatics. IEEE Press, New York (2006)
21. Marinakis, Y., Marinaki, M.: A Hybrid Honey Bees Mating Optimization Algorithm for the Probabilistic Travelling Salesman Problem. In: 11th IEEE Congress on Evolutionary Computation (CEC 2009). IEEE Press, New York (2009)
22. Haddad, O.B., Afshar, A., Marino, M.A.: Honey-Bees Mating Optimization (HBMO) Algorithm: A New Heuristic Approach for Water Resources Optimization. Water Resources Management 20(5), 661–680 (2006)

Artificial Negative Selection: Searching for an Appropriate Application Scenario

Yevgen Nebesov

Institute for Systems Engineering, Leibniz Universität Hannover
nebesov@sim.uni-hannover.de

Abstract. Despite numerous theoretical investigations on Artificial negative selection (ANS), there are still no useful scenarios in which this paradigm would outperform mainstream machine learning, statistical classification methods or other bio-inspired classification approaches. The aim of this paper is to identify main characteristics and requirements of a useful ANS scenario. Our investigations on this question led us to the need to extend the original ANS model proposed by Forrest et al. in [4]. The motivation of our work relies on the observation that biological mechanisms are not isolated mechanisms with a broad application range. They are only suitable for highly specific tasks and they might only be efficient in interaction with the rest of the biological environment.

1 Introduction

The Biological immune system (BIS) is a complex system of biological structures and processes allowing for protecting the host against disease by identifying and killing pathogens and tumor cells. The BIS is made up of two arms: innate and adaptive immunity. T cells are one of the key immune cells of adaptive immunity. They are subject to *negative selection* where T cells that match too strongly any self cell, a building unit of its host, are removed. The aim of negative selection is to guarantee that only T cells capable of matching non-self cells are released into circulation. An incomplete repertoire of self-cells presented during negative selection can lead to self-reactive T cells, i.e. to autoimmune reactions (false positives).

Due to its unique role in the BIS, negative selection has attracted a lot of attention from the research community. In 1994, the first model of Artificial negative selection (ANS) was presented by Forrest et al. [4]. The goal of this model is to generate *detectors* reflecting the capabilities of T cells. This model consists of the following two phases:

a) The detector generation phase, in which detectors are randomly generated and subsequently matched against each object in a self set. Only detectors that do not match any object from the self set are applied in the next phase.

b) The matching phase, in which detectors are applied to classify objects. Objects that match any detector do not belong to the self set.

E. Hart et al. (Eds.): BIONETICS 2011, LNICST 103, pp. 230–235, 2012.

D'haeseleer et al. [2] proposed several detector generation schemes for objects represented as binary strings. To match binary strings, he applied the r-contiguous bits (rcb) matching rule [8]. Balthrop et al.[1] introduced a generalization of the rcb matching rule called r-chunk matching rule. Ji and Dasgupta [6] investigated negative selection in real-valued vectors. Stibor et al. [9] reviewed several negative selection algorithms for binary and real-valued spaces and compared their classification performance with one-class Support Vector Machines. A recent theoretical result [3] shows that generation of string-based detectors, including binary strings, can be done in linear time.

Despite these efforts, a useful ANS scenario in which ANS could outperform other classification approaches has not been proposed yet. As the results of Stibor et al. [9] suggest one can expect that classical machine learning or statistical classification approaches are superior to negative selection. We see some methodological reasons for this fact. First, in the BIS, negative selection is not an isolated mechanism. On the contrary, it is a part of a complex chain of mechanisms interacting with each other when detecting a pathogen. Hence when looking for an ANS scenario, it might also be necessary to model the interaction with other bio-inspired mechanisms. The second reason is that negative selection operates under certain environmental and task-specific conditions. For example, during negative selection a complete or nearly complete repertoire of self cells is presented. T cells produced by negative selection are thus expected to provide a high pathogen detection rate. For this reason, we do not consider assessment of detection rate and false positive rate for an ANS scenario to be a central task. Another environmental condition, or rather restriction, is the distribution of T cells throughout the host. This makes an exhaustive simultaneous pathogen testing by the complete T cell repertoire challenging.

Any ANS scenario should reflect the characteristics of the scenario in consideration.

Unlike in [2,8,1,6,9,3], our focus stays on investigating the interdependence between the environmental characteristics of a potential ANS scenario and the scenario itself. We restrain from investigating any new detector generation schemes, detector types or classification results. Our ambition is to derive a feasible ANS scenario in which this classification approach can dominate other classification approaches. Our goal is to specify the ANS paradigm using its distinctive characteristics and features.

This paper is organized as follows. The next section introduces an Artificial Negative Selection Paradigm. Section 3 discusses its particular components and finally Section 4 concludes the paper.

2   Artificial Negative Selection

2.1   Notation

We apply the following notation:

1) U is the universe of all possible samples
2) $S \subset U$ is the self-set representing the total set of self cells in the human body
3) S_0 is the available training subset of the self-set $S_0 \subseteq S$, which represents a set of self cells collected in the thymus.
4) $N := U \setminus S$ is the potential non-self set.
5) $\mathbb{T}$ is the system's time axis
6) $N_0(t) \subset N$ is the set of the already seen non-self instances at the time t.
7) D is the set of detectors playing the role of the T cells
8) $\forall d \in D$ is a characteristic function $d : U \to \{0, 1\}$ that checks whether a tested sample belongs to the predefined subset, covered by this detector $U_d := d^{-1}(1) \subset U$. For generalization, we also mention fuzzy detectors which are modeled as fuzzy characteristic functions $d : U \to [0, 1]$ [5].
9) $U_D := \bigcup_{d \in D} U_d$ is the total covering by D.
10) $\rho(y, k) : N \times \mathbb{N} \to \mathbb{R}_{>0}$ is the average time or complexity of the response to the appearance of a non-self instance y after it was already detected k times.
11) $\Phi(t) : \mathbb{T} \to U \cup \{\emptyset\}$ is the event function describing the appearance of the new instances. For example, $\Phi(3sec.) = x$ means: a sample x has appeared at the time stamp 3sec after the system launch.
12) $v : \mathcal{P}(U) \to \mathbb{R}_{\geq 0}$ is a volume measure to estimate the size of a subset in U. For example, if U is discrete, v is the number of elements in a subset. If $U = \mathbb{R}^2$, then v is the area of a subset.

2.2 Characteristics of ANS Paradigm

We refer ANS paradigm to the category of on-line one-class classification schemes, operating under the following conditions (C1-C3) and tasks (T1-T3) an:

C1 $S_0 \approx S$: There are about $10^3 - 10^4$ peptides presented in the human thymus, which are typically 8-11 amino acids long [7]. This is enough to cover all possible self peptides of the host (the human proteome consists of circa 10^7 amino acids). Thus, we treat the presence of a complete self set during the detector generation as a conditional requirement of any proper ANS application.

C2 $\forall d \in D \ v(d) \ll v(N)$: Each particular detector can cover only a small area of potential non-self instances. This feature makes negative selection distinctive from the classifiers such as neural networks or support networks machines. In the case of these classification approaches, in order to establish a class membership of a tested instance, one has to evaluate a single function. Such a function can be treated as characterization for a whole class. The detectors instead can characterize only some small subset of a class. This granular structure has an advantage for a possible adaptation because a re-learning of SVM implies a re-building of the entire classification function, whereas ANS offers a possibility to make changes local on the detectors.

C3 $\forall t_0 \in \mathbb{T} \ \forall y_0 \in \tilde{N}(t_0) \ \forall y \in N \setminus \tilde{N}(t_0) \ \exists t' \in \mathbb{R} \cup \{+\infty\} \ : \ \forall t \in [t_0, t'] \ \mathbb{P}(\Phi(t) = y_0) > \mathbb{P}(\Phi(t) = y)$: This is an environmental feature and at the same time condition of negative selection. It supposes, that each non-self instance has relative to other non-self instances an increased probability to reappear in some time period after it was detected.

T1 $U_D \cap S \approx \emptyset$: The detectors are supposed to be self-tolerant in BIS. A small imperfection rate might not affect the functionality of BIS[7].

T2 $U_D \approx N$: BIS involves a generation of a large number of diverse detectors to cover all possible non-self instances[7].

T3 $k > j \Rightarrow \rho(y, k) \leq \rho(y, j) \; \forall y \in N$: This is a requirement to survive under the condition C3 by making use of already seen non-self samples in order to respond to their reappearance as quick as possible. This tasks means the more times a given non-self instance has already been detected the more quickly an organism can response if it will appear again.

The presented characteristics of ANS paradigm sufficiently restricts the applicability area for potential ANS based method. For example, the presence of a complete self class during the training is an uncommon case. However, in the next section we discuss several aspects of negative selection in BIS in order to find useful scenario settings for ANS applications.

3   Discussion

3.1   Random vs. Deterministic Detector Generation

The first step of the negative selection process is a generation of the detector set D. In Nature, T cells are randomly generated in the thymus and also randomly tested against the collected self samples[7]. In the artificial environment, randomness is not the only possibility to produce detectors. A deterministic approach may yield a competitive alternative. In the case of deterministic generation, the detectors are constructed by analyzing the available training self set S_0. This generation type seems to be more effective if the training set S_0 is given explicitly. But it is impossible to produce detectors deterministic if S_0 is implicit. An example of an implicit S_0 is a distributed data set with access restrictions. A number of banks could have a distributed data base containing information about credit card transactions and other user data. Each part of the base cannot share the information directly but can respond if a given detector covers the samples in this part (see Fig.1).

3.2   Matching of New Instances

The matching step is a processing step of ANS in which the detectors conduct the classification routine of newly appearing instances. Intuitively we want to test a new sample by the whole detector set D (see Algorithm 1).

This algorithm helps by identifying a proper environments for ANS applications. ANS operates with detectors covering N. Let us assume, we use alternative detectors covering S, so $\forall x \in S \exists d \in D : d(x) = 1$. This means that every

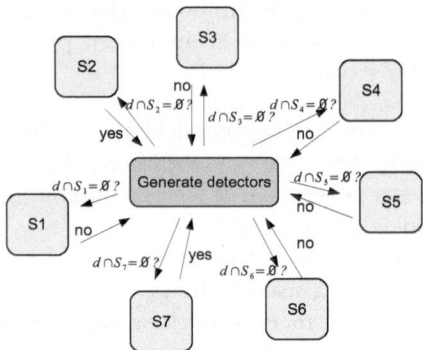

Fig. 1. An implicit training set $S_0 = \bigcup\limits_{i=1}^{7} S_i$

Algorithm 1 Matching of new instances procedure

---

input: x {a new sample}
input: D {the detector set}
for all $d \in D$ **do**
  if d(x) = 1 **then**
    return $x \in N$ {x is non-self}
    break;
  end if
end for
return $x \in S$ {x is self}

---

non-self instance should undergo testing by the whole detector set to establish its foreignness (see Algorithm 1). In BIS, this may not be possible since detectors are distributed throughout the body. This observation has led us to the possible beneficial scenario characteristics of the ANS - the distributed systems with limited rights and responsibilities of the components. Other characteristics are security concerns: each particular detector covering a subset of N does not share the information about S. In the opposite case, the detector would disclose this information.

Now let us assume, that this matching procedure operates under conditions C1-C3 to fulfill the tasks T1-T3. And let us, especially concentrate on the task 3). The way, how this classification scheme behave adaptive regarding the reappearance of an already known non-self instances, depends only on the order of detectors applied for matching. Since the condition $d(x) = 1$ breaks the algorithm, one might try to position the detectors in order to break the procedure as soon as possible. This observation can be used to fulfill the task of ANS's accelerated response to the reappearance of already known instances. However, the way how to reorder the detectors depends directly on the concrete application's scenario and can not be provided in the general case. The fuzzy detectors are more interesting for this adaptation method then the "yes/no" detectors because they bring more maneuvering place by allowing an adjustment of their fuzzy matching functions.

4   Conclusion

Design of immune inspired mechanism requires a careful study on the "working conditions" of their biological origins. Previous ANS research was biased primarily by not taking into account scenario features of designed applications. In this paper we stated that process design must rely on the given artificial environment and defined required characteristic for a set of proper scenarios. However, these characteristics sufficiently restrict an application area for ANS. We also stated that treating ANS as a classifier misinterprets its nature. This affects the way how we assess ANS applications. Much more important question regarding the quality of a given application is not its detection rate or false positives rate, but its ability to be adaptive in terms of changing environment. Understanding the advantages of ANS paradigm we pointed out beneficial scenario characteristics for a potentially useful ANS application. These are security concerns restricting disclosure of the self set and the use of fuzzy detectors.

Acknowledgement. This work was supported by the German Research Foundation (DFG) under the grant no. SZ 51/24-3 (Survivable Ad Hoc Networks – SANE).

References

1. Balthrop, J., Esponda, F., Forrest, S., Glickman, M.: Coverage and generalization in an artificial immune system. In: Proceedings of the Genetic and Evolutionary Computation Conference, pp. 3–10 (2002)
2. D'haeseleer, P., Forrest, S., Helman, P.: An immunological approach to change detection: algorithms, analysis, and implications. In: Proceedings of the 1996 IEEE Symposium on Computer Security and Privacy (1996)
3. Elberfeld, M., Textor, J.: Efficient Algorithms for String-Based Negative Selection. In: Andrews, P.S., Timmis, J., Owens, N.D.L., Aickelin, U., Hart, E., Hone, A., Tyrrell, A.M. (eds.) ICARIS 2009. LNCS, vol. 5666, pp. 109–121. Springer, Heidelberg (2009)
4. Forrest, S., Perelson, A.S., Allen, L., Cherukuri, R.: Self-nonself discrimination in a computer. In: Proceedings of the 1994 IEEE Symposium on Research in Security and Privacy, pp. 202–212 (1994)
5. Gonzalez, F.A.: A study of artificial immune systems applied to anomaly detection. PhD thesis (2003)
6. Ji, Z., Dasgupta, D.: Real-Valued Negative Selection Algorithm with Variable-Sized Detectors. In: Deb, K., Tari, Z. (eds.) GECCO 2004. LNCS, vol. 3102, pp. 287–298. Springer, Heidelberg (2004)
7. Košmrlj, A., Chakraborty, A., Kardar, M., Shakhnovich, E.I.: Thymic selection of t-cell receptors as an extreme value problem (2009)
8. Percus, Percus: String matching for the novice. AMM 101 (1994)
9. Stibor, T., Mohr, P.H., Timmis, J., Eckert, C.: Is negative selection appropriate for anomaly detection? In: GECCO, pp. 321–328 (2005)

Spatio-temporal Modeling and Simulation of *Mycobacterium* Pathogenesis Using Petri Nets

Rafael V. Carvalho[1], Willem Davids[1], Annemarie H. Meijer[2], and Fons J. Verbeek[1]

[1] Imaging & BioInformatics, Leiden Institute of Advanced Computer Science
Leiden University, Niels Bohrweg 1, 2333 CA Leiden, The Netherlands
{carvalho,fverbeek}@liacs.nl, s0775045@umail.leidenuniv.nl
[2] Molecular Cell Biology, Institute of Biology
Leiden University, Einsteinweg 55, 2333 CC Leiden, The Netherlands
a.h.meijer@biology.leidenuniv.nl

Abstract. Computational modeling of biological systems is becoming increasingly important in the endeavors to better understand complex biological behavior. It enables researchers to perform computerized simulations using a systems biology approach, in order to understand the underlying mechanisms of certain biological phenomena. It provides an opportunity to perform experiments that are otherwise impractical or infeasible *in vivo/vitro* experiments. In our approach we propose to model and simulate the pathogenesis of *Mycobacterium marinum* using Petri Net formalism based on data obtained from analysis of microscope images and to provide a three dimensional visualization of the whole infection process and granuloma formation. Image analysis will provide an accurate estimation of the infection in a structured database which will be used for the construction of the Petri Net model. The results of the simulation and analysis of the infection behavior will be visualized in 3D.

Keywords: Formal methods, Petri nets, biological system, simulation.

1 Introduction

Understanding the behavior of biological systems has been a key challenge for scientists. Experimental research often requires a complex tuning of *in vitro* and *in vivo* experiments, demanding high costs in terms of time and resources. To manage these efforts biologists have used computational modeling in support of a better understanding of complex biological behavior.

The necessity of process and visualize such processes has prompted biologists to harness computers to construct, verify and validate models that describe biological systems. In this way various computational formalisms have been used to explain the mechanism behind a biological system in intuitive and easily analyzable terms. Among these, there are methods inspired by biological phenomena such as Brane Calculi [5], P Systems [17], Biocham [4] and Calculus of Looping Sequence (CLS) [2], and also computational methods based on concurrent systems such as Calculus of Communicating Systems (CCS) [7], π-Calculus[19] and Petri nets [6, 9, 18]. In their

E. Hart et al. (Eds.): BIONETICS 2011, LNICST 103, pp. 236–241, 2012.
© Institute for Computer Sciences, Social Informatics and Telecommunications Engineering 2012

work [8], Fisher and Henzinger present a comprehensive discussion about mathematical and computational methods and distinguish them by their different approach in representation of biological phenomena.

Segovia-Juarez et al. [20] use an agent-based model (ABM) of granuloma formation during *Mycobacterium tuberculosis* infection, simulating the granuloma containment and dissemination to with a 2D tool. ABM is a class of computational methods that is used for simulation of the actions and interactions of autonomous agents on simultaneous operations and interactions based on rules that govern the simulation. In their approach Segovia-Juarez et al. conclude that an agent-based model is an appropriate tool for exploring complex spatio-temporal system such as granuloma formation. However, they did not provide the verification and validation of their simulation model.

The verification and validation of a simulation model is extremely important in a simulation process so as to provide results as close as possible to reality that is simulated. Verification involves debugging the model to ensure that it works correctly; whereas validation ensures that you have built the right model. Petri nets have been utilized in biology as a modeling and simulation instrument to create, verify and validate models that represent biological phenomena.

Ivan Mura and Attila Csikász-Nagy[15] define a Petri net model of the cell cycle of a budding yeast cell. They use the Möbius tool to validate their Petri net model and compare the result of the simulation with the model solved by ODEs and experimental observations in the literature. They simulate the budding yeast cell cycle using the Petri net formalism presenting their results on a Cartesian graph instead of a visualization tool.

In this project we intend to develop a system for modeling, simulating and visualizing the process of *Mycobacterium* infection and granuloma formation. We will integrate information obtained from image analysis with a Petri net formalism to create, verify and validate the simulation model and visualize the results with a 3D rendering tool. The verification and validation will be covered by Petri net tools that simulate and check the model according to input information from image analysis data. This paper gives an overview of the approach, starting from the image analysis to generate the data input, passing it through to the Petri nets networks modeling the simulation, after which it will be visualized in a 3D tool. The paper intends to provide the basic- layout and addresses the modeling challenges; at this point formal definitions or results are not included.

2 Infection of *Mycobacterium Marinum* in Zebrafish

Human Tuberculosis is caused by the bacteria *Mycobacterium tuberculosis*, and this infection causes around 1.8 million deaths a year [22]. The bacteria are capable of evading the human immune system in different ways. A closely related species called *Mycobacterium marinum* (*Mm*) can be used to study pathogenic mycobacterial infection in animal models, such as the zebrafish [21]. After infection, zebrafish immune cells, mostly macrophages, migrate to the site of infection and take up the bacteria by a process called phagocytosis. Inside the macrophage non-pathogenic bacteria will be broken down in a lysosome, a specialized compartment containing anti-bacterial compounds. However, similar to *M. tuberculosis, Mycobacterium marinum* escapes from lysosomal degradation and infects the macrophage itself. This severely compromises the immune system, and the bacteria multiply inside of the infected macrophage, over time causing its death. Other macrophages get attracted to the infection, and take up the infected macrophages and bacteria. The process repeats itself, these macrophages get infected too by the bacteria. In this way aggregates of immune cells, containing the bacteria but are unable to get rid of them, are formed. As these aggregates grow, structures develop that are referred to as tuberculous granulomas, lumps of immune cells that surround the infection [3, 12]. These granulomas are the hallmark of the tuberculosis diseases in human and animal models.

Assessment of the infection process and granuloma formation ranges from the epidemiology to genetic levels. In order to understand this process, it is necessary to model the behavior of this disease on a multiple scales; from molecular, cellular, to the tissue level scale. Mariano et al. [14] review recent modeling efforts in capturing the immune response to *Mycobacterium tuberculosis* (*Mtb*) emphasizing the importance of a multiscale modeling approach. In our case, we intend to elaborate this idea to develop a systems biology approach of the *Mm* infection integrating both modeling and experimental aspects with a visualization tool to present the information of the infection and granuloma formation in an interactive manner to allow the exploration, analysis and understanding of the infection process.

3   Image Analysis Input

To serve as input for the model, different measurements will be done on the basis of confocal laser scanning microscopy (CLSM) images of zebrafish embryos infected with *Mm* and the following granuloma formation. Both the zebrafish embryos and bacteria come from transgenic lines that express fluorophores making it possible to look at the individual immune and bacterial cells in the zebrafish. Zebrafish embryos will be injected with *Mycobacterium marinum* at the 8 hours post fertilization (hpf) stage. At this point the embryos have developed a functional innate immune system which will initiate an immune response, leading to the formation of granulomas [12]. The infection will be traced in time, from the moment of infection to up to six days later.

Images of infected zebrafish from 1 dpf up to 6 dpf are acquired. In images the change over time are tracked to get measureable parameters, i.e. intracellular/ extracellular bacterial growth rate, bacterial spread and migration rates, growth and position rate of granulomas. These rates are obtained from automated image analysis which consist in recognizing the patterns by analyzing the specific fluorescent signal in the images. Nezhinsky and Verbeek [16] have presented an approach in which individual zebrafish embryos are recognized and counted in images with multiple object instances and in varying orientations. The recognition results in a mask that is used in the images analysis. We can use this approach to recognize the infection sites and analyze the size, amount and the concentration of an infection for each image and for each stage of infection. A statistical analysis will be performed to analyze and classify these data providing a probabilistic infection data structure that can be modeled further using a Petri net.

4   Using Petri Nets to Modeling Biological Systems

The Petri net (PN) formalism, a computational method for modeling concurrent systems, has been providing a natural and promising modeling technique used to model behavior in biological systems [6]. Reddy et al. [18] presented a method of representation for metabolic pathways, and Hofestädt [11], expanded Reddy's method to model metabolic networks; they were the pioneers in using Petri nets to model biological phenomena.

Standard PN models are discrete and non-temporized (time is implicit, the marking graph accounts for the possible sequence of event). This classical method is very useful for the modeling of reactions that do not include time or probability.

To model a more complex system, extensions methodology of the classical Petri net provide additional possibilities in modeling; i.e. in Colored Petri Nets (CPNs) data values are assigned using different colors as data structure [13]; Stochastic Petri nets (SPNs) probabilities are added to the different choices for the transitions to make these suitable for representing networks of biochemical transformations and signaling pathways [15]. In addition, there are also other extensions such as Hybrid Petri nets (HPNs) and Hybrid Functional Petri Nets (HFPNs) allowing for coexistence of both continuous and discrete process[1].

Taking advantage of the modularity property of Petri nets, we define different networks that will represent each scale – molecular, cellular and tissue – model and connect them using hierarchical nets, i.e., creating subnets for each scale using the CPNs or SPNs formalism. High level processes dealing with the organism and its organ systems, and lower level processes, dealing with, for instance, gene expression patterns and cell-to-cell signaling, can be incorporated into the same model.

Currently, we are using Snoopy [10] with Charlie as tool set; these tools are available for building and analyzing PNs networks. We have chosen to use a Colored Petri Net to model the early stages of the infection process and granuloma formation. This consists of macrophages phagocytosis of mycobacteria, migration of infected macrophages, growth of mycobacteria inside of individual macrophages and the aggregation and granuloma maturation.

5   3D Visualization Tool

The visualization tool will provide a 3D visualization environment including a GUI, a set of functions to configure the simulation and a program structure that supports flexible development of the system. Graphical toolkits (i.e. OSG, GLUT, VTK, etc.) are required to build the interactive 3D rendering application, it consists of functions and commands to specify objects and operations on a 3D scene.

The tool needs to interact with the Petri net model obtaining all information checked by the PN tool in order to provide a 3D animation of the infection process and granuloma formation. Therefore, it is necessary to make a data interface between the visualization tool and the model reading through an output file of the snoopy tool.

All components (Image analysis, PN and 3D module) will be interconnected as a simulation system which can be viewed by the user solely in terms of its input, output and transfer characteristics without any knowledge of its internal mode of operation. In Figure 1 the structure of the system is depicted.

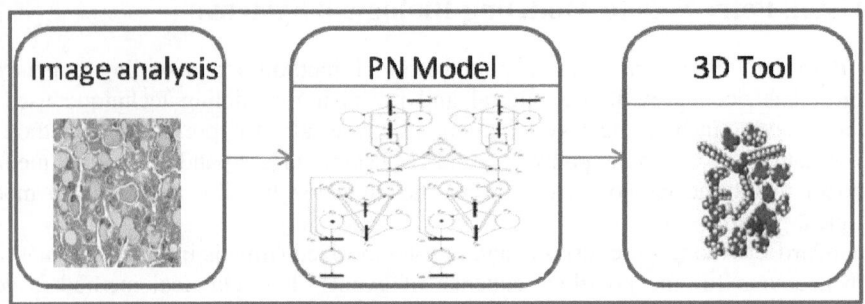

Fig. 1. Structure of the system with the interconnection modules. Figure merely illustrative.

6 Concluding Remarks

Using a systems biology approach, integrating modeling, experimental and visualization aspect is essential to create a computational system that reflects biological behavior, providing an opportunity to perform experiments that are impractical or not feasible *in vivo* or *in vitro* as well as generate testable hypotheses. Collaboration between researches from experimental and modeling background is crucial to achieve our goal of developing a system that will contribute to studies of infection and granuloma formation.

References

1. Alla, H., David, R.: Continuous and hybrid Petri nets. Journal of Circuits Systems and Computers 8, 159–188 (1998)
2. Barbuti, R., Caravagna, G., Maggiolo–Schettini, A., Milazzo, P., Pardini, G.: The Calculus of Looping Sequences. In: Bernardo, M., Degano, P., Zavattaro, G. (eds.) SFM 2008. LNCS, vol. 5016, pp. 387–423. Springer, Heidelberg (2008)
3. Bouley, D.M., et al.: Dynamic nature of host-pathogen interactions in Mycobacterium marinum granulomas. Infection and Immunity 69(12), 7820 (2001)
4. Calzone, L., et al.: BIOCHAM: an environment for modeling biological systems and formalizing experimental knowledge. Bioinformatics 22(14), 1805–1807 (2006)
5. Cardelli, L.: Brane Calculi - Interactions of Biological Membranes. In: Danos, V., Schachter, V. (eds.) CMSB 2004. LNCS (LNBI), vol. 3082, pp. 257–278. Springer, Heidelberg (2005)
6. Chaouiya, C.: Petri net modelling of biological networks. Briefings in Bioinformatics 8(4), 210–219 (2007)
7. Danos, V., Krivine, J.: Formal Molecular Biology Done in CCS-R. Electronic Notes in Theoretical Computer Science 180(3), 31–49 (2007)
8. Fisher, J., Henzinger, T.A.: Executable cell biology. Nature Biotechnology 25(11), 1239–1249 (2007)
9. Hardy, S., Robillard, P.N.: Modeling and simulation of molecular biology systems using Petri nets: modeling goals of various approaches. Journal of Bioinformatics and Computational Biology 2(4), 595–613 (2004)

10. Heiner, M., et al.: Snoopy-A Tool to Design and Execute Graph-Based Formalisms. Petri Net Newsletter 74, 8–22 (2008)

11. Hofestädt, R.: A Petri net application to model metabolic processes. Systems Analysis Modelling Simulation 16(2), 113–122 (1994)

12. Lesley, R., Ramakrishnan, L.: Insights into early *mycobacterial* pathogenesis from the zebrafish. Current Opinion in Microbiology 11(3), 277–283 (2008)

13. Liu, F., et al.: Computation of Enabled Transition Instances for Colored Petri Nets. In: Proc. AWPN, pp. 51–65 (2010)

14. Marino, S., et al.: A multifaceted approach to modeling the immune response in tuberculosis. Wiley Interdisciplinary Reviews. Systems Biology and Medicine 3(4), 479–489 (2011)

15. Mura, I., Csikász-Nagy, A.: Stochastic Petri Net extension of a yeast cell cycle model. Journal of Theoretical Biology 254(4), 850–860 (2008)

16. Nezhinsky, A., et al.: Pattern Recognition in Bioinformatics. Springer, Heidelberg (2010)

17. Pun, G.: A guide to membrane computing. Theoretical Computer Science 287(1), 73–100 (2002)

18. Reddy, V.N., et al.: Petri net representations in metabolic pathways. In: Proceedings of International Conference on Intelligent Systems for Molecular Biology, ISMB, vol. 1(115), pp. 328–336 (1993)

19. Regev, A., et al.: Representation and simulation of biochemical processes using the pi-calculus process algebra (2001)

20. Segovia-Juarez, J.L., et al.: Identifying control mechanisms of granuloma formation during M. tuberculosis infection using an agent-based model. Journal of Theoretical Biology 231(3), 357–376 (2004)

21. Stinear, T.P., et al.: Insights from the complete genome sequence of Mycobacterium marinum on the evolution of Mycobacterium tuberculosis. Genome Research 18(5), 729–741 (2008)

22. World Health Organization: Global tuberculosis control: epidemiology, strategy, financing, Geneva (2009)

Immune Inspired Adaptive Information Filtering: Focusing on Profile Adaptation

Nurulhuda Firdaus Mohd Azmi[1,3], Fiona Polack[1], and Jon Timmis[1,2]

[1] Dept. of Computer Science, University of York, YO10 5DD, UK
{huda,fiona,jtimmis}@cs.york.ac.uk
[2] Dept. of Electronics, University of York, YO10 5DD, UK
[3] Advanced Informatics School (AIS), Universiti Teknologi Malaysia (UTM)
International Campus, Kuala Lumpur, Malaysia

Abstract. This paper explores approaches to Adaptive Information Filtering (AIF) in the context of changing user interests. Based on the existing artificial immune system for email classification (AISEC), we demonstrate an effective extension to classification based on the body of emails. Widening this to the problem of AIF on dynamic web content, we propose to explore dynamic clonal selection algorithms (DCSAs) that include dynamically changing thresholds.

Keywords: Artificial Immune Systems, Adaptive Information Filtering, Profile Adaptation, AISEC, dynamic clonal selection.

1 Introduction

In adaptive information filtering (AIF), the relevance of the data is determined in accordance with the changing (adaptive) needs of a particular user [15,9]. The ambiguity in representing a document, the uncertain nature of a user's information needs and the associated formulation process [8] contribute to the difficulties encountered in creating a user profile. A user profile should dynamically adapt to drifts in users' interest and 'learn' with the changing interests. Inaccuracies in user profile affect the quality of recommendation. These challenges of profile adaptation to information filtering pose an interesting and challenging research area.

User interests change over time, driven by changes in the user's environment and knowledge. Dynamic profile adaptation is an example of multi modal dynamic optimization, (MDO) [10]. Our research is exploring use of artificial immune system (AIS) algorithms such as the dynamic clonal selection algorithms (DCSA) for profile adaptation. Some additional features of profile adaptation are as follows.

1. A user is interested in many topics at once, and a topic of interest may consist of related subtopics. This means that a dynamic user profile needs to represent multiple topics of interest.
2. A system supporting dynamic user profiling needs to maintain and adapt a diverse population of profiles in parallel.

E. Hart et al. (Eds.): BIONETICS 2011, LNICST 103, pp. 242–247, 2012.

3. User involvement is crucial because continual relevance feedback is the basis for the updating and fitness evaluation of profiles. However, user feedback may be unstable for many reasons. A user may not be very discriminating, or may have wide, shallow interests.

Through profile adaptation, the profile becomes open to its environment with the addition and removal of 'topics'. The profile constantly changes structure in response to changes in the stream of user feedback to documents. This paper summarises work on the existing AISEC algorithm for profile adaptation (Section 2). Our extensions to AISEC are shown to perform at least as well as Secker's algorithm [13] on a large corpus of emails (Section 3). We then presents our proposals for a new adaptive user profiling approach based on the DCSA (Section 4) – the work is based on a meta-probe analysis [14] of the adaptive user profiling problem and AIS algorithms [1,2], which showed that immune-inspired characteristics such as degeneracy and pleiotropism were of potential value in such contexts. Section 5 concludes our work.

2   Using AISEC for Interest Classification

The task of email classification is widely used as an experimental platform for exploring the effect of changing user interests. Secker et al [13] have developed an artificial immune system for email classification (AISEC) which classifies emails as interesting or uninteresting according to the subject and sender of the email. The AISEC algorithm removes uninteresting email from a user's inbox. The system has been shown to be capable of continuous learning, adapting to changes in the user's interests.

The AISEC system[13] uses inspiration from the behaviour of B-cells in the immune system to remove uninteresting emails from the system. In the immune system, there is a set of naive B-cells and a set of memory B-cells; when a naive B cell meets an antigen, it is stimulated, and becomes a memory B cell. In the algorithm, B cells have a feature vector. The feature vector is populated with words from the email subject and sender fields – the training phase of the algorithm populates the feature vector of naive B cells. As a result, memory B cells represent examples of words from uninteresting e-mails. After training, the algorithm processes emails as they arrive, treating the email as an antigen. The words extracted from the email subject and sender fields are compared to the feature vectors of existing memory B cells. The degree of matching is calculated as an *affinity measure*, which is compared to the preset *affinity threshold*. If the B cell affinity exceeds the threshold, then this B cell is said to recognise that email antigen – it is a candidate uninteresting email. At this point, user feedback is needed to determine whether the email is in fact uninteresting. A confirmation results in a reward to the memory B cell, and this results in cloning of this memory B cell. Cloning produces another B cell which may be mutated, according to predefined mutation rates. Mutation involves replacement of one word from the feature vector. Finally, in order to prevent unlimited growth in the population of memory B cells, a cell-death process is implemented in which cells which have not received sufficient stimulation over a period of time are purged from the system. Because the algorithm handles a stream of emails, and regular user feedback, the algorithm is dynamic.

Prattipati and Hart [12] have modified the AISEC algorithm to improve the speed of adaptation and the overall accuracy of the classification algorithm. However, AISEC use has been limited to the email subject and sender fields.

The previous AISEC algorithms [13,12] extract words only from the email subject and sender fields. This limits the potential diversity of the *gene library* (the set of all words from feature vectors of memory B cells that recognise uninteresting emails). This in turn reduces the diversity generated by cloning and mutation, since, when mutation is performed, a word from this library replaces a word from a cell's feature vector. In order to recognise new topics of interest and removal of existing topics of interest, we need a large library of words, and we need to be able to detect synonyms. There are several ways to increase the diversity of words in the gene library. Our first modification to AISEC is to consider only the body of the email, as this provides a richer set of words. In addition, we use the WordNet corpus as a source of synonyms. This allows more accurate classification of emails, and improves the capability to identify new and potentially uninteresting e-mail.

3   Comparing Performance of AISEC Versions

To compare performance, we run Secker's [13] original algorithm and our version (which uses email bodies and the WordNet corpus) on a large corpus of real-world email messages from the Enron Corporation[1]. The data consists of over 600,000 emails from the email accounts of 158 employees, and was released in 2004. Although the dataset is large, many users' folders are sparsely distributed. For the comparison experiment, a subset of 100 folders, containing 2420 messages was used. The system was trained with a training set of 200 emails.

The parameters for the algorithm are as for Secker's algorithm, but the values have been calculated by empirical testing with this dataset[2].

To evaluate the classification, we calculate the true and false positive and negative classification rates. We then use traditional *classification accuracy*[3], and Mathew's Correlation Coefficient (MCC) [3]. MCC was chosen because it is unaffected by sampling biases, which may occur when the dimensions of the learning sets are very different.

First, we test the (alternative) hypothesis that the median accuracy of the two algorithms is not statistically (rank sum) and scientifically (A test) different. We have carried a statistical analysis based on the non-parametric Rank-Sum statistic, and the A test to compare medians.

---

[1] http://www.cs.cmu.edu/~enron/

[2] The parameters are: affinity threshold (Ka) = 0.5, level of stimulated memory B cell (Ksm) = 80, level of stimulated naive B cells (Ksb) = 175, classification threshold (Kc) = 0.3, constant cloning rate (Kl) = 3.0, constant mutation rate (Km) = 0.3, training threshold (Kt) = 20. For a full explanation see [13].

[3] Ratio of correctly classified instances to the total number of instances in the test set.

Table 1. Statistical comparison of median accuracy of original and extended AISEC

rank-sum: p-value 0.0100
A test: A value   0.62099

The result of the statistical test is shown in Table 1. The p-value indicates that, with 95% confidence, the two algorithms are statistically and scientifically different. However, the A value shows that the difference has only a small effect. The predictive accuracy and MCC (on a scale 0.0 to 1.0) is shown in Table 2. The summary statistics show that the extended AISEC, operating on a much larger corpus of words, performs at least as well as Secker's original algorithm on this dataset.

Table 2. Median predictive Accuracy and MCC value for classification using extended AISEC and original AISEC

	Accuracy	MCC
Extended AISEC	90.10%	0.60
Original AISEC	86.67%	0.54

The analysis has only looked at the B cells' ability to reject uninteresting email. The plan is to carry out further analysis of the extended version of AISEC, to determine its ability to classify emails into multiple categories, whilst adapting to changing user interest, and to attempt to determine the maximum number of categories that can be effectively classified.

4   Proposing a DCSA for Profile Adaptation

AISEC variants perform well on email classification. However, we are interested in AIF for a wider range of documents, for instance, web content.

Various other adaptive immune inspirations have been proposed. Dynamic clonal selection (DCS) has been identified as an AIS algorithm that supports learning in dynamically changing environment [6,7,5,4]. In DCSAs, adaptation is improved by evolution of the gene library as well as clonal selection and hypermutation of the population of memory B cells, referred to as *detectors*. The AISEC extension has shown that large documents can be classified by the existing cloning and mutation of memory B cells. To evolve the extended-AISEC gene library would require evolution of the WordNet corpus. To do this effectively is a non-trivial research exercise.

Nanas et al [11] propose an approach to profile adaptation [11] using threshold tuning, whilst [9] proposes adaptation of the affinity threshold; this has not been implemented. The research uses an immune-network metaphor to represent user interests, which is rather different to the approach used in AISEC. In our proposed work, DCS is used to maintain user interest profiles with a gene libraries maintaining the set of terms that can be added to the profile during mutation. We also propose to incorporate adaptation of the affinity threshold through learning. Some issues that arise in the design of the algorithm are as follows.

- **Dynamic Evolution of Self**
  In immune-inspired systems, an underlying principle is to remove antigens that are not *self*. In adaptive systems, we need to be able to re-define self over time, by the addition of new detectors and removal of ones that are no longer useful.
- **Change in Thresholds.** In AISEC, memory B cells repeatedly match incoming emails and the maturation of memory B cells occurs when the predefined affinity threshold is reached. Dynamic adaptation suggests that the affinity threshold should be able to change as user interests change. To accommodate this, we propose to implement two thresholds. An immature memory B cell is continually stimulated by new word matches until a fixed *activation threshold* is reached; it then becomes a mature memory B cell. The mature memory B cell continues to be stimulated by new word matches, and the affinity threshold is used to determine whether it has been used often enough to remain in the population. Adaptation occurs because the affinity threshold changes with changing user interests – and all mature memory B cells are continually rechecked against the changing affinity threshold.
- **Lifespan of Mature Detectors.** Thresholding results in turnover of the mature detectors, which have a finite lifetime. A key issue is to ensure that some representation of memory B cells receiving low stimulation remains in the population. This makes it easier to recognise changed interest in previously-categorised topics.
- **Treatment of user feedback.** In maintaining the population of mature memory B cells, feedback from the user is interpreted as a *co-stimulation signal*. We need to ensure that a B cell can be stimulated both by word matches and by positive user feedback.

5   Conclusion

The paper explores approaches to AIF in the context of changing user interests. We demonstrate that email classification approaches that have been shown to be effective can be extended from consideration only of email header and subject, to consideration of the body of emails – a much less constrained word set. We have also shown the use of synonyms (from WordNet) in maintaining diversity of terms. Further analysis is needed on the extended AISEC algorithm. This includes sensitivity analysis of the algorithm's performance against minor changes of parameter values.

We propose to explore DCSAs that extend this work to include dynamically changing thresholds as well as evolution of the gene library.

References

1. Azmi, N.F.M., Timmis, J., Polack, F.: Profile adaptation in adaptive information filtering: An immune inspired approach. In: Proceedings of the IEEE Int. Conf. on Soft Computing and Pattern Recognition (2009)
2. Azmi, N.F.M., Timmis, J., Polack, F.: Towards a principled design of bio-inspired solutions to adaptive information filtering. In: Proceedings of the IEEE Int. Conf. on Engineering of Complex Computer Systems, ICECCS (2010)
3. Baldi, P., Brunak, S., Chauvin, Y., Andersen, C.A.F., Nielsen, H.: Assessing the accuracy of prediction algorithms for classification: An overview. Bioinformatics, 412–424 (2000)
4. Cayzer, S., Smith, J.: Gene Libraries: Coverage, Efficiency and Diversity. In: Bersini, H., Carneiro, J. (eds.) ICARIS 2006. LNCS, vol. 4163, pp. 136–149. Springer, Heidelberg (2006)
5. Cayzer, S., Smith, J., Marshall, J.A.R., Kovacs, T.: What Have Gene Libraries Done for AIS? In: Jacob, C., Pilat, M.L., Bentley, P.J., Timmis, J.I. (eds.) ICARIS 2005. LNCS, vol. 3627, pp. 86–99. Springer, Heidelberg (2005)
6. Kim, J., Bentley, P.J.: A model of gene libraries evolution in the dynamic clonal selection algorithm. In: Bentley, P.J., Timmis, J. (eds.) Int. Conf. on Artificial Immune Systems, pp. 182–189 (2002)
7. Kim, J., Bentley, P.J.: Immune memory and gene library evolution in the dynamic clonal selection algorithm. Genetic Programming and Evolvable Machines 5, 361–391 (2004)
8. Mostafa, J., Mukhopadhyay, S., Lam, W., Palakal, M.: A multilevel approach to intelligent information filtering: Model, system and evaluation. ACM Transaction on Information Systems 15(4), 368–399 (1997)
9. Nanas, N.: Towards Nootropia: A Non-Linear Approach to Adaptive Filtering. PhD thesis, The Open University (2003)
10. Nanas, N., De Roeck, A.: Multimodal Dynamic Optimization: From Evolutionary Algorithms to Artificial Immune Systems. In: de Castro, L.N., Von Zuben, F.J., Knidel, H. (eds.) ICARIS 2007. LNCS, vol. 4628, pp. 13–24. Springer, Heidelberg (2007)
11. Nanas, N., De Roeck, A., Uren, V.S.: Immune-Inspired Adaptive Information Filtering. In: Bersini, H., Carneiro, J. (eds.) ICARIS 2006. LNCS, vol. 4163, pp. 418–431. Springer, Heidelberg (2006)
12. Prattipati, N., Hart, E.: Evaluation and Extension of the AISEC Email Classification System. In: Bentley, P.J., Lee, D., Jung, S. (eds.) ICARIS 2008. LNCS, vol. 5132, pp. 154–165. Springer, Heidelberg (2008)
13. Secker, A., Freitas, A.A., Timmis, J.: AISEC: an artificial immune system for e-mail classification. Evolutionary Computation 1, 131–138 (2003)
14. Stepney, S., Smith, R., Timmis, J., Tyrrell, A., Neal, M., Hone, A.: Conceptual frameworks for artificial immune systems. Int. J. Unconventional Computing 1(3), 315–338 (2006)
15. Tauritz, D.R., Sprinkhuizen-Kuyper, I.G.: Adaptive Information Filtering Algorithms. In: Hand, D.J., Kok, J.N., Berthold, M. (eds.) IDA 1999. LNCS, vol. 1642, pp. 513–524. Springer, Heidelberg (1999)

Asynchronous Idiotypic Network Simulator

Kevin Sim

Edinburgh Napier University, Edinburgh, EH10 5DT, UK
k.sim@napier.ac.uk

Abstract. This paper describes improvements in both the efficiency and scalability of a computer simulation originally developed to illustrate idiotypic network theory in the immune system. The original synchronous model was streamlined using software engineering principles to improve efficiency before being adapted to function as a multi threaded application using asynchronous communication that better mimics the decentralised nature of the the biological system that it demonstrates.

Keywords: Artificial immune systems, Idiotypic network theory.

1 Introduction

This paper describes improvements in efficiency and scalability of an Artificial Immune System (AIS) simulation [1] originally developed to illustrate Idiotypic Network Theory (INT) and adapted in [4] and subsequent publications to investigate the affect of altering the shapes of antibodies recognition areas on the topologies of idiotypic networks. A brief summary of the biology is given before describing improvements made to the model, concluding with a statement as to the usefulness of such simulations in the domain of immuno-engineering.

2 Biological Inspiration

The vertebrate immune system is a complex biological system whose prime function is to protect its host against potentially harmful foreign bodies (pathogen). It is composed of a number of interacting cells of which B-Lymphocytes or B-Cells are responsible detecting and eliminating invading pathogen such as bacteria, viruses and other particles that are deemed potentially hazardous.

The human immune system is thought to be capable of detecting any shaped *antigen* using a repertoire of 10^{12} lymphocytes that each produce around $10^5 - 10^7$ identical antibodies. An antibody marks a pathogen for destruction through a process of binding in which a region on antibody known as the *paratope* physically binds with an area on the antigen called the *epitope*. A lock and key analogy is frequently used to describe the binding process in which the strength of the binding (*affinity* is proportional to the degree of complementarity between the two shapes.

Figure 1 shows a diagram of a B-Cell with one of its antibodies enlarged to highlight its paratope that in figure 2 bonds with an invading antigens epitope.

E. Hart et al. (Eds.): BIONETICS 2011, LNICST 103, pp. 248–251, 2012.

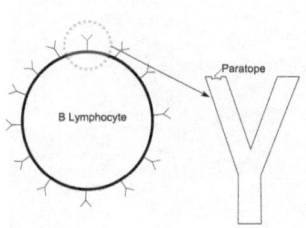 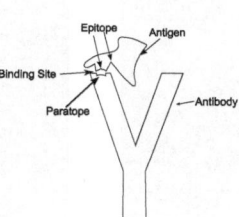 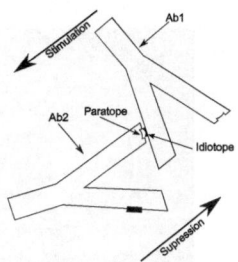

Fig. 1. A B-Cell Showing an Enlarged Antibody

Fig. 2. An Antigen recognised by an Antibody

Fig. 3. Idiotypic Stimulation and Suppression

In the classical immune model, called the *self recognition view*, antibodies bind only to *antigen*. Binding results in a cloning process and mutation process, resulting in a proliferation of antibodies that better fit the antigen. This view was challenged when Jerne[5] proposed his *self assertive* idiotypic network theory which asserts that antibodies can also recognise other antibodies, creating a network of stimulatory and suppressive signals which can be sustained in the absence of antigen, and was claimed to both determine the immune response and explain immune memory. Figure 3 shows antibody Ab_1 recognising an area, dubbed an idiotope, on Ab_2 along with the resultant stimulatory and suppressive signals that characterise an idiotypic network. Although Jerne originally described these interactions as asymmetric, subsequent research has favoured a symmetrical network in which there is no distinction between the recognising and recognised cell.

AIS based on INT[1] have been successfully used for optimisation, robot control and pattern recognition[2]. The following section describes a simulator developed originally to illustrate INT[1] and later adapted to investigate the affect of *shape space*[2] on the resultant topology of an idiotypic network[4].

3   Idiotypic Network Simulator

In the simulation presented in [4] antibodies are added one at a time at a random position on a 100×100 grid. Given an antibody, Ab_1, as labelled in figure 4 the complementary matching area is defined as a circle centred at the point geometrically opposite the centre of the grid radiating for a predetermined distance with the influence exerted by cells in this region on Ab_1 reducing linearly from the centre to zero at the circumference. Figure 5 shows the system after a number of iterations with one antibody, its area of affinity and stimulating cells highlighted.

The total stimulation, S_j, detected by an antibody, A_j, is calculated by summing the stimulation received from all cells situated within its complementary

---

[1] The original Idiotypic Network inspired AIS was presented by Farmer[3] in 1985.

[2] The shape space is the area of recognition which for most AIS is assumed spherical.

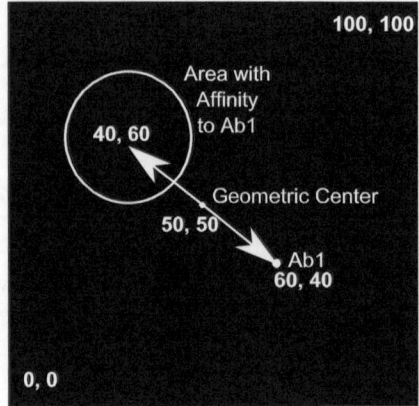

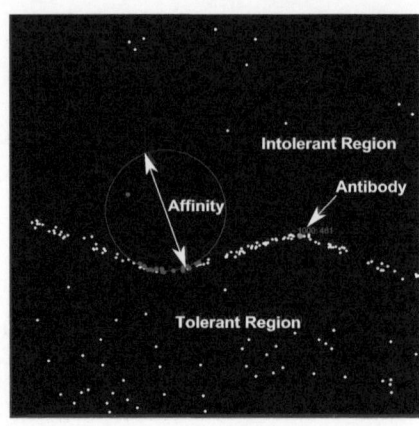

Fig. 4. Cell to Cell Affinity in Complementary Regions

Fig. 5. Antibodies naturally divide the space into antigen tolerant and intolerant regions

area of recognition using equation 1 where C_i is the concentration of antibody A_i, n is the number of antibodies in the system and D_{ij} is the Euclidean distance from the circumference of the complementary region to antibody A_i.

$$S_j = \sum_{i=1}^{n} C_i D_{ij} \tag{1}$$

If the stimulation an antibody receives is between an upper and lower threshold then its concentration is increased otherwise it is decreased with antibodies with a concentration of zero removed from the system. By varying the initial concentration, radius and thresholds for stimulation different networks emerge. The simulation from which the model was inspired[4] required that all antibodies check all other antibodies each iteration in order to determine their stimulation levels requiring, for a system with n antibodies, $n \times n$ calculations. This was improved upon in the following way. Each iteration an antibody, A_i, is added at a random location and antibodies with an affinity are added to to a list used to calculate stimulation so that in future iterations only those antibodies need be polled with the process reciprocated so that all antibodies with affinity become aware of A_i.

Whilst this improved the efficiency of the model by a factor approaching n it was also observed that the synchronous and centralised control of the system bore little resemblance to the biological system from which it was inspired which, from an engineering perspective, has exploitable features characterised by its autonomous decentralised, parallel and asynchronous nature.

A second asynchronous model was implemented with the following changes. A centralised thread introduces a new antibody, A_i, each iteration and attaches an event listener to all antibodies that have an affinity, with the process reciprocated as before. Each antibody executes on its own thread, firing an event every cycle which is responded to by those antibodies within its complementary matching area. Upon receiving an event, the transmitting antibody is polled by the receiving antibody which increments its stimulation accordingly. A delay is imposed on the central thread to limit the rate of cell introduction with another added to each antibody thread to allow sufficient time to accumulate stimulation which is decremented using a decay rate each iteration. In order to maintain a network the system has to balance the delays imposed with both the decay rate and the original parameters.

3.1 Conclusions and Future Work

The simulation described benefits from decreased computational effort and increased scalability due to its distributed nature and is more consistent with the biological system it aims to model. The CPU time required, once a network is established, can be minimised by increasing the delay on threads without affecting network behaviour which remained faithful to that of the original simulation.

The simulation provides a tool for both illustrating and investigating the properties of idiotypic networks. The simulation enables the engineer to abstract properties of the biologiocal system and study the relevance of key parameters. In this respect, the tool contributes to the domain of *immuno-engineering* , defined by [6] as the abstraction of immuno-ecological and immuno-informatics principles, and their adaptation and application to engineered artefacts so as to provide these artefacts with properties analogous to those provided to organisms by their natural immune systems. This approach advocates the use of models such as the one discussed in this paper which take into account the differences between artificial systems and biological systems and enables extraction of minimal properties of a system required to achieve specified functionality.

References

1. Bersini, H.: Self-assertion versus self-recognition: A tribute to francisco varela. In: Proceedings of ICARIS, Canterbury (2002)
2. DasGupta, D.: Artficial Immune Systems and Their Applications. Springer-Verlag New York, Inc. (1998)
3. Farmer, J.D., Packard, N.H., Perelson, A.S.: The immune system, adaptation, and machine learning. Phys. D 2(1-3), 187–204 (1986)
4. Hart, E., Ross, P.: The impact of the shape of antibody recognition regions on the emergence of idiotypic networks. International Journal of Unconventional Computing 1, 281–314 (2005)
5. Jerne, N.K.: Towards a network theory of the immune system. Ann. Immunol. (Paris) 125C(1-2), 373–389 (1974)
6. Timmis, J., Hart, E., Hone, A., Neal, M., Robins, A., Stepney, S., Tyrrell, A.: Immuno-engineering. In: Hinchey, M., Pagnoni, A., Rammig, F., Schmeck, H. (eds.) Biologically-Inspired Collaborative Computing. IFIP, vol. 268, pp. 3–17. Springer, Boston (2008)

Design and Modeling for Self-organizing Autonomic Systems

Paul L. Snyder and Giuseppe Valetto

Drexel University, Philadelphia PA 19104, USA

Abstract. Describing, understanding, and modeling the behavior of systems built upon self-organizing principles (such as many bio-inspired systems) is key to engineering self-organizing systems that can solve problems in real computing environments. Capturing the properties of the micro-macro linkage that connects local behaviors of system components to global emergent properties of the system as a whole is particularly important. Different kinds of models have been proposed, each focusing on a different aspect of the problem: descriptive models provide notations that support the design activity and the application of self-organzing principles; validation models allow formal examination of dynamic properties; and analytic models provide techniques for mathematical exploration of abstracted collective behaviors. Our goal is to identify and select the best tools available from these families, extend them where needed, and tie them together to support the creation and analysis of self-organized autonomic computing systems in an integrated way.

1 Introduction

Modern computing environments continue to expand in scale and complexity, challenging traditional approaches to engineering and control. These systems place increasing burdens on the limited resources of skilled administrators, calling for the development of new approaches to design, maintenance and management. *Autonomic computing* is one such paradigm, with the goal of developing adaptive software that can to respond to changing conditions and environments without human intervention.

Some branches of autonomic computing have used *exogenous control* approaches drawn from fields such as control theory. More recently, there has been growing interest in *endogenous control*, where desirable global system behaviors emerge from the *self-organizing* interactions of many individual elements. This latter approach is of particular interest as computing systems reach ultra-large scales [1]. In the face of huge numbers of components with heterogeneous properties and ownership, high levels of dynamism, and indistinct system boundaries, centralized command and control approaches to autonomic problems (such as self-configuration, optimization, and diagnosis) become impractical.

Studies of naturally occurring self-organization in complex adaptive systems (particularly from biology) have proven to be a rich source of inspiration for

E. Hart et al. (Eds.): BIONETICS 2011, LNICST 103, pp. 252–258, 2012.

building distributed computing systems along similar principles. Phenomena observed in natural systems are often useful as high-level metaphors, but specifications with additional levels of formality and detail are needed to turn them into working distributed systems. Experience shows that bio-inspired metaphors often lead to computing systems with extreme *scalability* and *robustness*. Since their self-organizing rules operate in local neighborhoods, an individual computing element in a self-organizing system need only consider a small subset of the total information available in the system at any given time; as a consequence, systems designed using these principles may be able to scale more easily to very large numbers of interacting elements. These systems also show emergent adaptive behaviors, leading to redundancy and self-healing even when the system is perturbed or disrupted in new and unpredictable ways.

Self-organization and emergence are related concepts. Discussions of self-organization highlight the radically decentralized nature of an algorithm, which can achieve increased order (either temporal, spatial, or functional) without external direction. Emergence refers to a two-way linkage between the lower (*micro*) level interactions between individual system elements, and higher (*macro*) level properties exhibited by the collective system behavior, which are not explicitly specified in—nor reducible to—those lower-level interactions [2].

As the interplay between the micro- and macro-level can be elusive, there is an inherent difficulty in trying to predict and explain the behavior of a self-organizing system. Indeed, subjective surprise on the part of an observer has been proposed as a test for the existence of emergence [3].

Designers of self-organizing systems that must operate in real-world circumstances have to ensure that the such systems satisfy requirements across a wide range of operating conditions. As such, describing, understanding, and modeling the nature and mechanics of the micro-macro linkage in a system is key to engineering self-organizing systems, in order to develop some guarantees (for instance, statistical) on global-level properties, and to show that the system is robust and reliable across a range of conditions.

One way to reason about such systems is through expressive modeling techniques. There has been much recent research in this area, but it very much remains an open problem. Different models capture different concerns and try to explain different aspects of the problems of engineering self-organization and emergence. Certain approaches provide descriptions that support the design of new systems. Others provide techniques to validate the self-organizing dynamics of a system and generate proofs about them. Others build mathematical models for the analysis of their behavior at some abstraction level.

Design, validation and analysis are all important concerns. However, it is currently quite difficult to connect models from these various families, and achieve a faithful and insightful representation of a self-organizing computing system. We maintain that a multi-perspective modeling process is needed, and our goal is to identify and select the best available modeling techniques from each of these families, and integrate them to better support the creation of new systems, as well as the analysis of existing ones.

2   Motivating Scenario

The need for a set of tools allowing reusable design, validation, and analysis has been highlighted by our previous experiences building self-organizing systems. Several of these systems have centered around *Myconet*, an unstructured overlay protocol for peer-to-peer networks [4]. The Myconet algorithm takes inspiration by fungal growth patterns, in order to build an efficient self-optimizing superpeer topology that can also rapidly self-heal in response to damage or attacks.

Myconet has proven to be flexible, and we have used it as a platform for other self-organizing applications, including load-balancing in distributed service networks [5], and attack detection and mitigation [6]. Those works have demonstrated the applicability of Myconet to a variety of self-organizing scenarios and applications. New applications can be added as a layer on top of the the the core Myconet; to leverage some of its properties, they may need to interact with the core and may induce alterations to the lower-level behaviors.

Each extension to Myconet brings additional insights into self-organizing dynamics, but also show us the limitations of ad hoc approaches to design and analysis of each new application. These experiences have led us to investigate formal tools and models. We have examined a number of models and tools from the literature. Our intent is to select chosen techniques to model Myconet and its extensions, leading to the development of an integrated set of useful tools and modeling approaches that can be reused for the principled design of other Myconet extensions, and at the same time enhance our understanding of self-organizing design in general.

3   Proposed Research

The design of self-organizing control mechanisms requires tools beyond those offered by traditional software engineering, and a large number of models have been proposed in the literature. These models have been developed separately for different purposes, and they may examine very different features of a system. We have classified the models we have examined into three general families: *descriptive models*, *validation models*, and *analytic models*. These model families—with some relationships among them—are depicted in Figure 1. Descriptive models are used to support the design of self-organizing systems. They draw on and extend tools both from the software engineering (such as UML diagrams and agent-oriented SE) and complex adaptive systems fields. Examples include several efforts to catalog *design patterns* for endogenous control, and arrange those patterns into composable ways [7]. Other research efforts have proposed descriptive notations (often as extensions to the Unified Modeling Language [UML]), using tools such information flow analysis [8], causal loop diagrams [9], or explicit modeling of a "meso-layer" [10], to explicitly bring self-organizing and emergent mechanisms into the design process. Most notations are only semi-formal, and as such may not provide much support for more rigorous analyses of system properties. A major use of descriptive models is to facilitate the actual development of

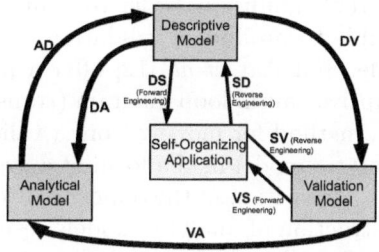

Fig. 1. Families of models for self-organized autonomic systems. Arrows represent feasible transitions, as discussed in Section 3.

a self-organizing application through a process of forward engineering (transition DS in Figure 1). The reverse engineering of a design model from a working self-organizing system (transition SD) is also possible; however, automated reverse engineering tools are not necessarily effective at highlighting the design elements that are significant to self-organization. In particular, automated identification of self-organizing patterns is an open problem.

The goal of validation models is to study the properties of a system, and verify that it fulfills its requirements. This topic has been studied extensively in software engineering, but the very complex interactions in self-organizing applications (which also tend to be highly distributed and highly concurrent), and their emergent nature, pose a great challenge to traditional validation approaches. Additional tools are needed in order to formally reason about the dynamics of such systems. Examples of proposed validation techniques include the use of formal languages, *e.g.* event calculus [11] and stochastic π-calculus (Sπ) [12]. Other approaches use established formalizations from software validation research and the analysis of biological systems, such as Stamatopoulou *et al.*'s proposed hybrid of Communicating X-Machines and Population P Systems [13]. Validation models can also serve as starting points for developing system (transition VS). Pitt *et al.* [11] take an event calculus description of an observed sociological system and compile it into an executable simulation that exhibits actual self-organizing dynamics. Direct derivation of a validation model from an actual system by reverse engineering (transition SV) may also be possible.

Analytical models offer another approach to understanding self-organizing systems by reducing their dynamics to a set of simplified mathematical equations that often abstract away lower-level details. In the best case, an analytic model allows terse descriptions of global behaviors and provides a foothold for the exploration of an application's parameter space. A number of analytical models use concepts from physics or information theory, and focus on characterizing the statistical complexity of the system, for example by measuring its entropy [14]. These approaches provide tools for assessing the self-organizing properties of a system once it has been developed, but are less helpful for initial system design.

In general, it is difficult to determine which model may be best for a particular system. Each modeling approach evidently has its own strong suit, but also carries some limitations. In some cases, is possible to translate between specific models belonging to different families in order to gain the benefits of each. For instance, it may be possible to produce a validation model starting from a descriptive one; for example, Sudeikat *et al.* [12] offer a procedure for deriving a (Sπ) model from a descriptive causal loop diagram (transition DV in the figure). Renz *et al.*, [9] propose a method for moving from a validation model (specified in Sπ calculus) to an analytic model in the form of a system of differential rate equations (transition VA). Transition in the opposite direction is hardly possible as the higher level of abstraction of analytic models results in loss of formal detail about the system. Moving directly from a descriptive to an analytical model may be possible, although the literature only suggests preliminary steps. For example, the specifications of self-organization design patterns may be decorated with analytic annotations about its behavior, as shown in [7]. Direct generation of a descriptive model from an analytic model (transition AD) is difficult, for the same reasons as transition AV: analytical models tend to abstract out a lot of detail. Since analytical models of biological phenomena (or other Complex Adaptive Systems) developed in the natural sciences are often used as the starting points for the design of new computing systems, it is typical to try to capture them by hand as thoroughly as possible within a suitable descriptive model.

While some works that demonstrate transitions between specific models in different families exist, they represent point solutions, and this problem is in general currently outstanding. Some recent research has also focused on how to move beyond an ad hoc approach to the design of self-organizing systems, and specify a methodology. Several design methodologies have been proposed, typically beginning with a descriptive model of a proposed system (*e.g.*, [15]); however, these methodologies do not directly address the roles of models from different families in their design process.

The issue of a more unified and comprehensive approach to the specification of self-organized systems is therefore an open area of investigation. The objective of our research is to understand how existing modeling techniques can be connected and integrated, when necessary by providing some extensions and bridges among them, so that we can solve in an integrated way the problems of sound construction, production of proofs and guarantees , and dynamics analysis, in the challenging domain of self-organized autonomic software systems. Such integration would not only represent an enhancement that may facilitate the work of researchers and practitioners involved in the development of those systems; it is also likely to lead to new insights on how different concerns interact and intertwine, and impact the creation of emergent mechanisms and self-organized dynamics, and analyze their efficacy.

4   Research Plan and Conclusions

All three types of models discussed in the previous section play an useful role in the engineering of self-organized autonomic systems, and research is needed to develop a unified approach to modeling.

In the initial stages of our research, we will select best-of-breed modeling techniques from each of the three families. We will apply these to our core

Myconet platform, in order to understand the benefits and limitations of each, and extend them where necessary. Based on our experiences, we will develop a preliminary integration of the selected models, resulting in a collection of transformations for moving between different models and a process for applying these. Next, we will extend the models to include the layered applications that have been developed on top of Myconet, and examine how they can be expressed as specializations of or extensions to the base Myconet models, while continuing to refine the integration. Finally, we intend to release Myconet and its models, along with the integrated collection of tools.

This research will result in insights into which of the large number of models proposed in the research literature are most useful and beneficial for designing and analyzing self-organizing systems in a unified way, as well as an integrated approach to applying those multiple modeling techniques. It will also provide a demonstration of a principled approach to the design of self-organizing applications using the Myconet platform, which will be instrumental in enabling its wider adoption. This unified modeling of Myconet may also serve as a blueprint or template of how the design and analysis of self-organizing applications in general can be addressed.

References

1. Northrop, L., Feiler, P., Gabriel, R.P., Goodenough, J., Linger, R., Longstaff, T., Kazman, R., Klein, M., Schmidt, D., Sullivan, K., et al.: Ultra-large-scale systems: The software challenge of the future. Software Engineering Institute (2006)
2. De Wolf, T., Holvoet, T.: Emergence Versus Self-Organisation: Different Concepts but Promising When Combined. In: Brueckner, S.A., Di Marzo Serugendo, G., Karageorgos, A., Nagpal, R. (eds.) ESOA 2005. LNCS (LNAI), vol. 3464, pp. 1–15. Springer, Heidelberg (2005)
3. Ronald, E.M.A., Sipper, M., Capcarrère, M.S.: Design, observation, surprise! A test of emergence. Artificial Life 5(3), 225–239 (1999)
4. Snyder, P.L., Greenstadt, R., Valetto, G.: Myconet: A fungi-inspired model for superpeer-based peer-to-peer overlay topologies. In: SASO 2009, pp. 40–50 (2009)
5. Valetto, G., Snyder, P.L., Dubuois, D.J., Di Nitto, E., Calcavecchia, N.M.: A self-organized load-balancing algorithm for overlay-based decentralized service networks. In: SASO 2011 (2011)
6. Snyder, P., Osmanlioglu, Y., Valetto, G.: Biologically Inspired Attack Detection in Superpeer-Based P2P Overlay Networks. In: Hart, E., et al. (eds.) BIONETICS. LNICST, vol. 103, pp. 99–114. Springer, Heidelberg (2012)
7. Fernandez-Marquez, J.L., Arcos, J.L., Di Marzo Serugendo, G., Casadei, M.: Description and composition of bio-inspired design patterns: the gossip case. In: 2011 8th IEEE International Conference and Workshops on Engineering of Autonomic and Autonomous Systems (EASe), pp. 87–96. IEEE (2011)
8. De Wolf, T., Holvoet, T.: Using uml 2 activity diagrams to design information ows and feedback-loops in self-organising emergent systems. In: Proceedings of the Second International Workshop on Engineering Emergence in Decentralised Autonomic Systems (EEDAS 2007),, pp. 52–61 (2007)
9. Renz, W., Sudeikat, J.: Modeling Feedback within MAS: A Systemic Approach to Organizational Dynamics. In: Vouros, G., Artikis, A., Stathis, K., Pitt, J. (eds.) OAMAS 2008. LNCS, vol. 5368, pp. 72–89. Springer, Heidelberg (2009)

10. de Cerqueira Gatti, M.A., de Lucena, C.J.P.: A bio-inspired representation model for engineering self-organizing emergent systems
11. Pitt, J., Schaumeier, J., Artikis, A.: The axiomatisation of socio-economic principles for self-organising systems. In: Proc. of IEEE Conf. on Self-Adaptive and Self-Organizing Systems, SASO 2011 (2011)
12. Sudeikat, J., Renz, W.: A Systemic Approach to the Validation of Self–Organizing Dynamics within MAS. In: Luck, M., Gomez-Sanz, J.J. (eds.) AOSE 2008. LNCS, vol. 5386, pp. 31–45. Springer, Heidelberg (2009)
13. Stamatopoulou, I., Kefalas, P., Gheorghe, M.: Modelling the dynamic structure of biological state-based systems. BioSystems 87(2-3), 142–149 (2007)
14. Marinescu, D.C., Morrison, J.P., Yu, C., Norvik, C., Siegel, H.J.: A self-organization model for complex computing and communication systems. In: Second IEEE International Conference on Self-Adaptive and Self-Organizing Systems, SASO 2008, pp. 149–158. IEEE (2008)
15. De Wolf, T., Holvoet, T.: Towards a methodology for engineering self-organising emergent systems (2005)

Author Index